WATER, SANITATION, ENVIRONMENT AND DEVELOPMENT

WATER, SANITATION, ENVIRONMENT AND DEVELOPMENT

Selected papers of the 19th WEDC Conference
Accra, Ghana, 1993

Edited by John Pickford,
Peter Barker, Adrian Coad, Margaret Ince,
Rod Shaw, Brian Skinner and Michael Smith

Practical Action Publishing Ltd
27a Albert Street, Rugby, CV21 2SG, Warwickshire, UK
www.practicalactionpublishing.org

in association with

The Water, Engineering and Development Centre (WEDC)
Loughborough University of Technology
Leicestershire LE11 3TU, UK

© WEDC, Loughborough University of Technology, 1993, 1994

© Intermediate Technology Publications 1986

First published 1994 \Digitised 2008

Reprinted by Practical Action Publishing
Rugby, Warwickshire UK

ISBN 978 1 85339 240 5

All rights reserved. No part of this publication may be reprinted or reproduced or utilized in any form or by any electronic, mechanical, or other means, now known or hereafter invented, including photocopying and recording, or in any information storage or retrieval system, without the written permission of the publishers.

A catalogue record for this book is available from the British Library.

Since 1974, Practical Action Publishing has published and disseminated books and information in support of international development work throughout the world. Practical Action Publishing is a trading name of Practical Action Publishing Ltd (Company Reg. No. 1159018), the wholly owned publishing company of Practical Action. Practical Action Publishing trades only in support of its parent charity objectives and any profits are covenanted back to Practical Action (Charity Reg. No. 247257, Group VAT Registration No. 880 9924 76).

Conference Secretary: Rowena Steele
Papers typeset by Karen Betts and Kathy Brown
Designed by Rod Shaw

ACKNOWLEDGEMENTS

The editors would like to acknowledge the commitment
of the Local Organizing Committee whose arrangements helped to ensure
the success of the 19th WEDC Conference.

CONTENTS

INTRODUCTION		xi
Section 1	COMMUNITY MANAGEMENT	1
	D R Birch and P D H Field *UK/Pakistan* **Rural community-managed projects, Balochistan**	3
	Ms Rudith S King and Dr Romanus Dinye *Ghana* **Women, children, water/sanitation development**	7
	Dr T V Luong, F Njau and A Y Kahesa *Tanzania* **Towards self-management: water and sanitation**	9
	Pius B Mabuba *Tanzania* **Strategies for effective community involvement**	12
	Ofori MacCarthy and Dr Andrew Livingstone *Ghana* **Community planning of water supplies**	15
	Baby Mogane-Ramahotswa *S Africa* **The spread effect of a pilot community project**	18
	Ms Fati Mumuni *Ghana* **Working with rural folk in the Northern Region**	21
	Adrian Wilson *S Africa* **Community participation: Umgeni Water's approach**	24
Section 2	GROUNDWATER	29
	Robert R Bannerman *Ghana* **Siting of sanitary landfill and faecal treatment**	31
	NK Sekpey and S A Larmie *Ghana* **Nitrate pollution of groundwater sources at Oyarifa**	33
	Richard M Teeuw *UK* **Low-cost GIS for water resources**	36
Section 3	HEALTH AND DISEASE	39
	Dr Sam Bugri *Ghana* **Community-based surveillance in GWEP, Ghana**	41
	Dr Sandy Cairncross *Burkino Faso* **Guineaworm eradication – Is the target attainable?**	45
	Mrs Jemima A Dennis-Antwi *Ghana* **Participatory methods in hygiene communication**	49
	Susanne Niedrum *Rwanda* **The need for hygiene education**	52

Section 4	**INSTITUTIONAL DEVELOPMENT**	55
	Peter Barker *WEDC* **Pricing water to recover costs**	57
	Erich Baumann *Switzerland* **Private sector involvement**	60
	Dr Mrs V Hemalatha Devi *India* **Legal module for environmental protection**	64
	Duncan Morris *Cote d'Ivoire* **Thinking things through**	66
	Professor S Mustafa *Nigeria* **Improving water supply through privatization**	70
	Dr Wilfred Owen, Jr *Egypt* **Utility development: Cairo, Egypt**	73
	Mike Wood *Canada* **Privatization of rural water supply**	76
Section 5	**IRRIGATION**	79
	Lana Abu-Hijleh *Palestine* **Treated wastewater re-use in the Gaza Strip**	81
	S K Agodzo, J M Gowing & M A Adey *Ghana/UK* **Trickle irrigation using porous clay pots**	85
Section 6	**SANITATION**	87
	A H Abel and S V Dohrman *Malawi* **Makata pumpable VIP latrine block**	89
	Rob Burgess *S Africa* **Rural school sanitation pilot project**	93
	Shamsul Huda *India* **Subsidy: to what extent?**	96
	Dr Joyce Malombe *Kenya* **Sanitation and solid waste disposal in Malindi**	98
	Maria Muller, Jasper Kirango and Jaap Rijnsburger *Tanzania* **An alternative pit latrine emptying system**	101
	John Pickford *WEDC* **Low-cost sanitation research and GARNET**	104
	Martin Strauss *Switzerland* **Treatment of sludges from on-site sanitation**	108
Section 7	**SOLID WASTE MANAGEMENT**	115
	S Mansoor Ali, Adrian Coad and Andrew Cotton *WEDC* **Informal sector waste recycling**	117
	Inge Lardinois and Arnold van de Klundert *Netherlands* **Small enterprises for solid waste recycling**	120

Section 8 WATER QUALITY 123

Dr Margaret Ince and Miss T I Ojo *WEDC/Nigeria*
Pollution in Lagos Lagoon systems 125

Grace Rukure, Shungu Mtepo and Cornelius Mukandi *Zimbabwe*
Water quality in family wells 127

Section 9 WATER SOURCES 131

Siaw Awuah and John Addy *Ghana*
Earth dams for RWS in Northern Region 133

Dr Layi Egunjobi *Nigeria*
Rainwater-harvesting initiatives in Ekpoma, Nigeria 137

Section 10 WATER SUPPLY 141

Dr Manu N Kulkarni *India*
Pumps, people and payments 143

Peter Smith and A Mbaye *UK*
Pipeline extensions spread benefits 146

Section 11 WATER TREATMENT 149

Dr Christopher J Austin *The Gambia*
Chlorinating household water in The Gambia 151

Michael D Smith *WEDC*
Domestic solar disinfection for potable water 154

J P Sutherland, G K Folkard, M A Mtawali & W D Grant *UK*
Moringa oleifera at pilot/full scale 156

Martin Wegelin and Kolly Dorcoo *Switzerland/Ghana*
Water treatment in northern Ghana 158

INTRODUCTION

by John Pickford

WEDC CONFERENCES are held alternately in Africa and Asia. Previous African venues have been in Kenya, Malawi, Nigeria (twice), Tanzania and Zimbabwe. In Asia we have been to Calcutta, Hyderabad in India, Kathmandu, Kuala Lumpur, Madras and Singapore. The emphasis is always on practical 'real world' (Third World) situations. The majority of authors and participants are involved in the field as planners, designers, builders, operators and maintainers. Others are concerned with health aspects, rural development, economics, management, research and training. Most topics deal with water supply and sanitation, but we are also interested in irrigation, solid waste management and improvements to the physical infrastructure of rural and urban areas. Most emphasis is given to low-cost methods of improving quality of life and to communal methods of development.

Although WEDC Conferences follow a more-or-less similar pattern which has proved best over the years, each is distinctive. In part the special features of a Conference depend on its geographical location and in part they relate to the Local Organizing Committee (LOC) with whom WEDC collaborates. Some LOCs have been engineering institutions, others academic establishments.

The Accra Conference was fortunate to have a LOC with wide interests. The Chairman was the head of a research institute. Members work for government departments, consultants, international bodies and NGOs. The enthusiastic secretariat was provided by an international NGO which stresses community management. Consequently community matters featured strongly in the Conference, even when discussing technological issues.

In selecting the contents of this book we have tried to give a representative sample of all the ninety-odd papers presented at the Conference. So readers will find that the main concerns of 'Accra 1993' are reflected. Community management has a major role, but does not dominate to the exclusion of technological, health, scientific, environmental, organizational, institutional and economic matters that are essential for good development.

Water, Engineering and Development Centre, Loughborough University of Technology, 1994.

SECTION 1

COMMUNITY MANAGEMENT

Rural community-managed projects in Balochistan

D R Birch and P D H Field

THIS PAPER DEMONSTRATES a methodology unique and workable within the constraints encountered. At the same time the approach does include important elements that are considered fundamental in striving towards community managed projects.

As a result of extensive socio-economic surveys of the rural population and short comings with regard to ineffective governmental public health policy, the Balochistan Integrated Area Development (BIAD) was formed in 1980. The programme was initially considered a pilot attempt by the government to mobilize the rural people in the development of basic community services. As well as providing basic infrastructure, the BIAD programme is intended to provide a means of initiating community development.

The basic objective of the BIAD programme is to improve the health status and quality of life for the rural population of Balochistan. This is further defined into the strategic objectives of reducing the alarming rates of mortality and morbidity, providing a ready access to safe drinking water and the development of a low cost sanitation programme.

This paper concentrates on the range of effects the overall objectives, project strategy, resources and constraints have had in selecting appropriate schemes for the highly diverse communities in rural Balochistan. It also deals with the broader issues of community participation and draws away from the implications of government policy and associated logistics.

Community participation is an integral part of the BIAD development programme. This involves appropriate levels of beneficiary participation in the planning, design, implementation and operation and maintenance of development projects. Such participatory levels incorporate the knowledge and expertise of the beneficiaries in preparing project designs, planning and implementation.

Aspects of beneficiary participation have been totally ignored in the past though the beneficiaries were expected to operate and maintain the schemes once completed. Often this resulted in under-utilisation or failure of schemes either because these did not meet the beneficiaries perceived needs or the technologies introduced were inappropriate. Sometimes the technologies introduced were not maintainable by the beneficiaries. In many cases operation and maintenance of such schemes have since been taken over by the Government.

Within the context of community managed systems and the expected benefits of lower capital and running costs, appropriate and socially accepted technologies; and beneficiary care and maintenance of the facilities provided; the approach has involved a number of clearly defined stages as follows:

- Acceptance of the cluster
- Formation of a cluster committee
- Signing of a formal agreement
- Design considerations and participatory levels
- Construction phase
- Operation and Maintenance

Acceptance of the cluster

During initial discussions with the villagers the responsibilities of the intended beneficiaries were discussed in detail. Villages were only considered for inclusion in the programme if full agreement was reached, with regard to the proposed criteria for development and the basis for community participation.

It is common wisdom in development work that initially people are likely to accept most conditions to get development programmes underway in their villages but forget the partnership agreements once the implementation commences. As a cautioning measure a brochure written in Urdu, detailing the implementation strategies and illustrated with simple sketches, was distributed among the beneficiaries.

Selection of schemes was based on a priority order of needs qualified in terms of availability of the water resources per capita, access to potable water and present levels of sanitation. Schemes were also selected on an equitable basis through the Districts and according to the Governments Strategic Development Plan.

Formation of a Cluster Committee

The first obligation of the intended beneficiaries is to form a Cluster Committee. Typically a Cluster Committee includes; a chairman, a secretary, a treasurer and a number of members depending on the population size and the numbers of villages within a cluster. The Cluster Committee members nominated by the villagers were representative of the whole village. Where the Cluster comprised more than one village each village had at least one representative on the Cluster Committee.

The Cluster Committee was formed during the time the sociological surveys were conducted. The key factor to determine at this stage was the level of service to be provided in consultation with the intended beneficiaries. For the water supply all schemes, except those involving hand pumps, required a distribution network with supply points. The basic level of service was identified as community standpipes or community tanks, with an attached array of standpipes.

Although this level was found to be acceptable to the cluster groups of remote areas, villagers of the relatively more developed areas requested compound connections. Whilst the project initially considered such demands, the final decision was made once the quality and quantity of water was determined, according to the affordability of the beneficiaries and the cost effectiveness of the scheme. Where it was agreed to provide compound connections the beneficiaries had to pay a proportion of the additional cost above that of the basic level of service.

Signing of a formal agreement

The Cluster Committee then signed a formal agreement with the District Commissioner and BIAD for the scheme on behalf of the intended beneficiaries. This was completed prior to the start of the detailed design and only when the a Cluster Concept Report, based on the sociological, hydrological and hydro-geological surveys, was approved by BIAD and the donors. The agreement, presented in Urdu and available in English translation, set out in clear terms the rights and obligations of all parties signing and include the following items of responsibility:

- to provide land free of charge and arrange the legal transfer to BIAD for the sites of the water supply utilities;
- to allow the contractor to utilise locally available materials such as sand, gravel, water and other natural resources at no extra cost, nor to charge any other royalties to the contactor;
- to assist in agreed items of the construction;
- as appropriate, to provide pump operators and valve men to operate the system;
- to open a bank account and collect funds to maintain the account at a sufficient level to cover at least one months estimated O&M costs;
- to maintain the scheme in its entirety;
- where appropriate to pay for all electricity charges;
- where appropriate to pay for compound connections and required pipes; and
- to guarantee that the water will not be used for irrigation purposes.

The land must have been transferred over to BIAD or to the communal ownership of the Cluster members at the time of the signing of the scheme agreement. At this stage and where appropriate the Cluster Committee should have finalised an agreement for the installation of an electrical connection. BIAD was responsible for the cost of the connection and the extension of power lines where required.

Design considerations and participatory levels

Initial studies of the potential water resources included possible use of surface water and groundwater sources. At the stage of preparing the detailed design of the schemes the beneficiaries were involved as key informants and co-decision makers. The design team consulted the beneficiaries on all aspects of location and the design of the scheme components.

Sullage water often proved to be a problem, in both compounds and community areas, with the extent of the problem frequently dependent on the volume of water available and used by the community. Improvement of village water supplies can give rise to additional drainage problems and cause existing problems to become more extensive. Fortunately the arid nature of the area served to reduce the potential for drainage problems. The actual significance of the problem varied from village to village according to ground conditions, topography and village layout.

Water quality is a major constraint to the development of any water source. Surface water river and canal sources are frequently polluted particularly from wastes from upstream villages. They are also turbid and may be subject to high silt loads, especially at times of heavy rainfall and during floods. Such sources always require treatment and hence therefore become very difficult to design for community managed schemes. Such schemes were not included in the programme and remained as potential public health schemes to be operated and maintained by the Government.

Surface water sources such as mountain streams, springs. karezes and the upper reaches of some small irrigation canals are often clear of biological contamination. Where such sources can be protected from pollution and therefore treatment becomes unnecessary, community managed schemes become more viable. This is particularly the case for gravity fed schemes where initial capital costs and future recurrent costs are minimised. In some areas of the Province, groundwaters were found to be too saline for potable use, however water sources in the Province are generally neither biologically nor chemically polluted from either agriculture nor industry.

Another design consideration, which has fundamental ramifications on the defined approach and the concept of community managed schemes, was the diverse backgrounds of the varies ethnic tribal groups. Two main ethnic groupings live in Balochistan, the Pathans in the north and the Baluch and Brahui tribes in the south. Areas on the boundaries of the Province may have ethnic minorities of Sindhis and Punjabis. The differences between the structures of the major groups of tribes were carefully borne in mind while formulating the approach, design and development of schemes in the Province.

Upon completion of the designs, the designers held formal meetings with the Cluster Committee during which site visits were conducted and design components of the scheme were again reiterated and agreed physically. The Cluster Committee then signed a formal statement that they approved and fully understood the scheme layout and various components of the proposed development programme.

Whilst the designs were being prepared the sociologists discussed the need for recruiting staff to operate and maintain the scheme with the Cluster Committee. The Cluster Committee were encouraged to select candidates for training as pump operators, valve men, maintenance staff and accountants. The training was then organised during the construction phase.

Construction phase

It was during this phase that the beneficiaries role changed from that of assisting in the decision making to that of physical participation in the development of the works. This stage was seen as critical in developing the ownership concept within the beneficiary groups.

There are two ways in which beneficiaries contributed to the construction works; by paying a small proportion of the total costs and by contributing labour. In line with the level of service, the beneficiaries provided the labour for the excavation of all the secondary trenching within the boundary of the villages and payment for all service pipelines within compounds. The contractors were responsible for the construction of the remaining works comprising the source and the distribution system to the villages and the mains through each village. Under such an arrangement, this has had the advantage that the contractor does not have to rely completely on the comparatively slower progress made by the beneficiaries. It also means that the system built by the contractor retains its integrity as an operational water utility. This also allows for possible non compliance of the beneficiaries for certain sections of the secondary pipeline, without interrupting the overall pipe network system.

Where compound connections were requested the beneficiaries where expected to pay an agreed sum for the connection and for the additional piping in extending lines within compounds. For water schemes involving a lower level of service, such as well mounted hand pumps, the beneficiaries level of contribution involved the excavation of the well.

During the construction phase O&M manuals for the schemes were prepared and given to the Cluster Committee. The manuals were produced in Urdu and included various simple sketches to further explain the text. Training programmes were organised and provided to the staff nominated by the Cluster Committee.

On completion of the construction phase a formal meeting is then held with the Cluster Committee during which all elements of the works are inspected. Agreement is then reached as to any defects which require rectification. Following this a formal handing over ceremony is arranged at which BIAD formally hand over the running of the scheme to the Cluster Committee.

Operation and maintenance

The beneficiaries will take the full responsibility for the operating and maintaining the scheme. However, BIAD will continue to provide a follow up service. This will comprise maintenance engineers based at district level to whom the beneficiaries will be able to refer for advice. Any forms of major maintenance requirement will receive the full support of BIAD under a future support programme.

As part of the maintenance programme the BIAD engineers will make routine visits every 2 to 3 months. During such visits a joint inspection of the system will be made by the BIAD engineers and members of the Cluster Committee. Any defects will be noted and recommendations and advice will be given as to the remedial action required. BIAD will also monitor the quality of water supplied in the schemes.

Sanitation programme

Under the sanitation programme, each household have agreed to excavate the pits, construct the superstructure and provide the required unskilled labour for the construction of VIP or twin pit pour flush latrines.

Cluster development

Most village organisations have for the first time been organising the communities to undertake a communal development scheme within their villages. Members of the community have organised the collection of funds, often involving substantial sums in relative terms, and have been deposited into their own bank accounts in readiness for financing the running of their scheme. The funds have been collected from all benefitting households, unlike many development projects in the Province where contributions are made by the few wealthy families.

Many Cluster Committees have formed sub-committees to plan their contribution to the construction whilst some have also formed women's committees of users of the public health facilities provided.

Once the present schemes are completed BIAD will have an on going responsibility to the communities, as defined in the formal agreement. As a part of the objective to mobilise the rural people in the development of basic community services, under future phases, BIAD would encourage further development for the more successful schemes. Examples of further development would include the following:

- arranging for and paying a proportion of the costs of training male and female teachers or health workers, who are from the village, where villagers agree to pay their salaries once they return to the village;
- the construction and/or equipping of basic health centres or schools. Although this would be only done where staffing is guaranteed, since the problem of providing these basic services in most villages is less related to buildings and more a problem of staffing;
- minor improvements to flood protection and irrigation works;
- improvements to village roads, where this is important to the social and economic development of the communities.

Monitoring and evaluation

The programme to date has involved the generation of a number of community managed projects, the success of which can only be evaluated within the limits determined by the overall objectives of the programme, defined by certain indicators.

Monitoring and Evaluation will be an essential part of the BIAD support programme, in ensuring the full integrity and self-sustainability of the development projects. However unless a strong participative approach to the planning and development of the schemes is adopted, objectives will not be achieved.

Health and hygiene

Where water and sanitation programmes have included a hygiene education component, improvements in health and child survival have been comparatively high. Although basic hygiene education does not entail vast amounts of funds, particularly when compared to the overall capital costs of the water supply programme, they do have dramatic and favourable health consequences.

The initial stages of the BIAD programme included elements of an organised health and hygiene education programme. The programme started during the final stages of the scheme design and prior to the construction phase.

Women, children, water/sanitation development

Ms Rudith S King and Dr Romanus Dinye

HITHERTO, IN GHANA and other developing countries, a lot of projects aimed at providing water and sanitation facilities to rural communities have failed in the creation of lasting and reliable structures. Even costly installations have deteriorated due, in the main, to lack of maintenance and repair caused by poor interaction and limited involvement of the principal users – women and children – at the decision-making level.

A user participation approach if adopted would have actively involved the target groups in the planning, implementation, monitoring and evaluation of the water and sanitation projects. Involvement entails sensitivity to community's problems and commitment on the part of actors and supporting agents. It involves a gradual capacity built-up, utilization of own resources and enhanced self reliance.

The above notion underlies the strategy adopted for the Rural Water and Sanitation Project currently underway in the Volta Region of Ghana. It is undertaken by the Ghana Water and Sewage Corporation in collaboration with UNDP. The operationalization principle is that any community which endures a project from these outfits should form a Water and Sanitation Committee (WATSAN Committee). The tasks of such a committee include the maintenance and operation of the water and sanitation facilities bequeathed to them.

Women and children in water and sanitation project

Women and children in the rural areas are dubbed as key change agents in water and sanitation projects because they have to do with the household chores, childcare, family welfare, household cleanliness, collection and utilization of water. But it is known amongst other things that the effective and sustained utilization of improved water and sanitation facilities will ultimately lead to the improvement of the living conditions of the rural communities. In that capacity, they must be viewed as planners, operators and managers of water and sanitation systems at their community levels. Consequently, a requirement that should be fulfilled as the placement of at least two women on each WATSAN Committee.

In their dual capacity as users in the first instance and planners, operators and managers of water and sanitation systems in the second instance, it has become imperative to adequately train women and children against the background of improved water supply and utilization as well as enhanced community and household sanitation in the rural areas. Such a training in a case study in the Hohoe and Jasikan districts of the Volta Region have yielded beneficial results. Members of the WATSAN Committees in the two districts were tutored at a three-day residential workshop in Water and Sanitation Management. A post-ante assessment showed the relevance of good training with women at the centre of the communities as change agents.

The training methodology and results

First case study

The training was done in Ewe, the language of the beneficiaries. The response by way of participation during the three-day period was dramatic albeit the intensity of the workshop proceedings. Nevertheless, the training methods were, however, appropriate and equipments used adequate. The material employed included diagrams, charts, handouts, posters and flip charts. The subjects covered comprised health, water and sanitation. The trainers were skilled and worked with the trainees in smaller groups.

In addition to the provision of wells and KVIP latrines, the project makes sure that the focal people are given adequate training to enable them to make an impact on their community. Training therefore becomes very relevant in water and sanitation development. Occasionally there is a need for refresher training and monitoring of the focal people. It is very easy for women to put what they learn into practice in their homes and communities since they are responsible for fetching water.

In another case study at Moshie Zongo in Kumasi, where the UNDP in an attempt to find solution to urban sanitation problems adopted the same promotional and training strategies, but now focusing on men due to cultural barriers preventing the involvement of women.

In the evaluation process, employing the achievement level technique indicated an impact level of 87,7% implying a rather very good output. The training was very fruitful and participants on self-examination were satisfied with the conduct of the workshop. It had revealed to them many things they did not know previously. For instance they became aware of the changes of visiting public latrines without shoes on and the consequences of using leaves in water they fetch in order to stop it from pouring. Moreover, they no longer have direct contact with water at the sources at which they draw. Used or dirty water is no longer poured into sources of drinking water.

The container used in fetching of water from the water pot is no longer used in drinking water. Some of the women have got separate containers for fetching water and different ones for drinking. However, some of the women complained that they do not have money to buy separate cups/containers for fetching water and for drinking, even though they realise the need for that.

Though the period was so short to feel the full impact of the training, in few communities members of the WATSAN Committees have managed to organise the community to educate them about these basic things. Others have not been able to do them but they have succeeded in educating their immediate families and neighbours. This was the result of Case Study One.

The second case study

The second case study is an urban sanitation pilot project in Kumasi involving the provision of 100 Kumasi-ventilated Improved Pit (KVIP) latrines in a community known as Moshie Zongo. Again, based on the same strategy of demand-driven approach and community involvement, a committee was established at the community level known as the Community Sanitation Committee (CSC) which is to perform the same task as the WATSON Committee. The CSC went through the same training, using similar training equipments and materials. After the training the trainees were expected to train members of the community as to how to keep maintenance and operate their KVIPs as well as how to keep good hygiene.

Unfortunately for this project, the CSC was made up of all men and in the community they often met only the landlords when in actual fact, it is the women and children who maintain the facilities. Moshie Zongo is a Moslem community and their custom does not allow women to be in the forefront of events. As a result women and children could not be included in the CSC. The effect is that the impact of the training could not be felt by the key managers of the facilities i.e. the women. They are found leaving the privy rooms open, dropping used anal cleansing materials in the privy room, and spitting in the privy room. One can attribute all these to the fact that the main managers of the facilities were not involved in the initial training of the trainers, the CSC; the women and children testified for that by saying they were not aware of all those negative practices.

Conclusion

In both case studies, children were left out altogether. Probably because of the assumption that mothers would educate their children. But this could further be facilitated if school curricula for children would include elements of management and utilization of water and sanitation systems in their households and communities.

The essential lessons to be learnt from these cases are as follows:

i) gender variables must be considered in the planning and implementation, monitoring and evaluation of water and sanitation projects.
ii) training programmes for water and sanitation projects should be done simultaneously with the implementation of the projects.
iii) the key users of the facilities should participate in all the training programmes.
iv) training programmes must be done using very appropriate training methods and equipments.

It is only by full involvement of the key users of the facilities at community level will feasible and viable solutions to community problems be attained, leading to a gradual built-up of the capacity to utilise and rely on own resources.

References

1. King, R.S., *Integration of women and women's groups into the rural drinking water and sanitation project in the Volta Region* (training component). Training Network Centre, Civil Engineering Department, University of Science and Technology, Kumasi, Ghana, 1992.

2. Kumasi Metropolitan Assembly. Moshie Zongo private home KVIP latrine project 1992.

Towards self-management: water and sanitation

T V Luong, F Njau and A Y Kahesa

SAFE WATER FOR ALL by the year 1991 was the target set by the Government of Tanzania during the declaration of the twenty-year Water Programme in 1971. The main objective was, and still is, to provide adequate potable water for rural areas within 400 metres from each household in the Mainland. The twenty-year Water Programme has now been extended to the year 2002. The Government subsequently declared full commitment of the implementation of the objectives of the International Drinking Water Supply and Sanitation Decade (IDWSSD) and the development goals for year 2000 endorsed during the World Summit for Children held at the United Nations in September 1990 and the subsequent National Summit in June 1991. Two of those development goals are universal access to safe drinking water and to sanitary means of human excreta disposal.

The Ministry of Health launched a latrinisation campaign "Health for ALL" (MTU NI AFYA) in 1973 to counter the high incidence of faecal born diseases. A decree issued in 1974 requires that each household must have a latrine and use it hygienically. The Ministry of Health in 1982 set the target of achieving universal latrine coverage in rural households by year 1991.

At the end of the twenty-year Water Supply and Latrinisation programmes and of the United Nations Water and Sanitation Decade, only 46% of the rural population and 67% of the urban inhabitants have access to safe water respectively. However, many of the existing water schemes, reported at over 35%, are not functioning. Over the years, latrinisation campaign has gained momentum. Currently, about 85% and over 90% of the households in rural and urban areas respectively are having latrines near home and are using them.

Constraints

Since independence, provision of water services has been the exclusive responsibility of the Government. Water is free for all. Throughout the 20-year Water Programme, Government's financial allocation to the water sector was low with an average of 6.4% of the total government development budget. Furthermore, in real terms the financial input was diminishing due to the devaluation of the Tanzania Shilling and the decline of external support. As a consequence, among others, these have had two major implications; the gradual erosion of the potential for increased coverage and further weakening of the sector management and maintenance system.

Naturally, limitation of financial resources for sector development has a direct impact on low coverage. Major bottlenecks which have hindered the sector's progress include:-

(a) investments focused on high cost technology and the absence of a well defined sector strategy;
(b) inadequate and weak sector planning, management and monitoring;
(c) least involvement of communities in the overall planning, implementation and operation and maintenance rendered frequent breakdown and eventually non-functioning of many schemes; and
(d) lack of inter-ministry coordination and inadequate linkage of the water and sanitation sector with other development programmes such as health, education, women's development and communication.

Sector policy and guidelines

Recognizing these weaknesses, the Ministry of Water, Energy and Minerals (MWEM) in 1987 embarked on the formulation of a National Water Policy which was officially launched in November 1991. On the other hand, the Ministry of Health (MOH) in addition to the enforcement of existing legislation relating to the provision of sanitary facilities, has prepared environmental sanitation guidelines which was officially announced in February 1991. Strategies to facilitate better implementation of the guidelines was developed recently in March 1993. These initiatives by the government aim at accelerating sector coverage and services sustainability through cost recovery and cost sharing; application of low cost technologies; strengthening sector planning, monitoring and management as well as enhancing operation and maintenance of schemes at community level through the involvement of women, establishment of village water and sanitation committees and village water funds.

Actions

The Ministry of Water, Energy and Minerals and the Ministry of Health have initiated the following actions, guided by the National Water Policy and Sanitation Guidelines, aimed at achieving sustainable services and increasing coverage through community-based self-financing and self-management.

Establishment of sector monitoring system

The framework for an effective but simple participatory monitoring system has been jointly developed in 1991, as shown in Figure 1, by MWEM, MOH and the Ministry of Community Development Women and Children in collaboration with other concerned Ministries.

The sector monitoring system will facilitate immediate generation of information/data at the grassroot level and enabling analysis, assessment of situation and taking remedial actions within the community, District,

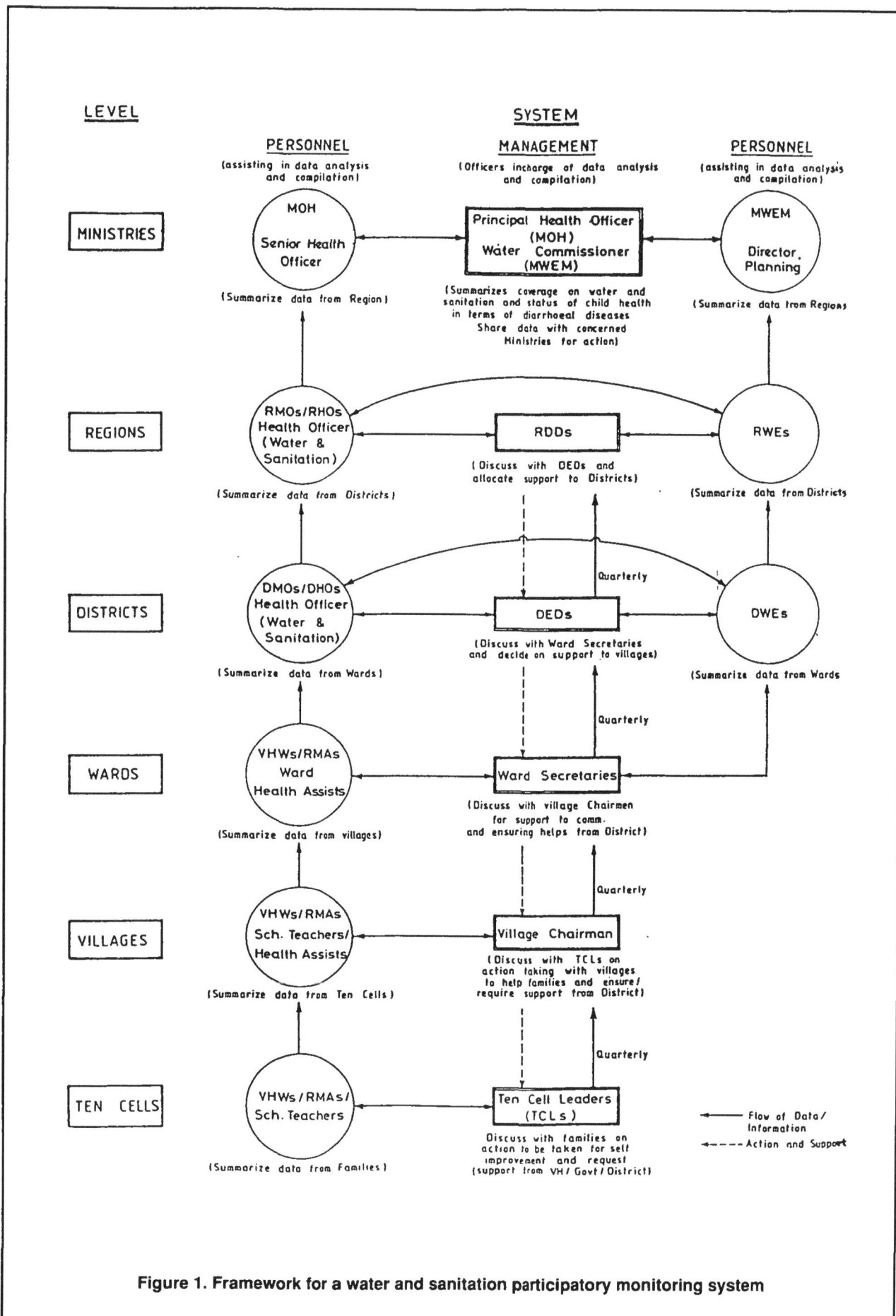

Figure 1. Framework for a water and sanitation participatory monitoring system

Region and Central levels. Constant feed back and regular follow-up by all concerned can ensure/pinpoint areas which required efforts of mobilisation or other support services. Sector monitoring units have already been established in the Division of Planning in MWEM and the Department of Preventive Services in MOH.

Implementation of national water policy

Strategies and Action Plan for implementation of National Water Policy have been developed by MWEM.

Actions have already been taken to initiate the transfer of one of the four major National Water Supply schemes to a Water Board or a Autonomous Body to manage the scheme through user-pay water tariff. Studies on aspects of operational and financial management of four urban water supplies revealed that water revenue earnings would be sufficient to run three out of the four urban water supplies. It is reported that many villages have already established village water and sanitation committees with 50% women members. Village Water Funds have been set up in many villages. System of support and distribution of spare parts to village level is yet to be initiated.

Low cost technologies such as shallow/medium deep wells installed with handpumps, small gravity feed schemes and rainwater harvesting are to be adopted wherever feasible.

Universal sanitation coverage

The relatively high latrine coverage in both rural and urban areas reflects that the majority of Tanzanians have already formed good habit of using latrines. Almost all households have built latrines entirely at their own cost. The Ministry of Health aims to achieve the universal sanitation coverage by year 1997. A National Plan of Action (1993-1997) has been developed focusing on the achievement of one latrine for every household and the promotion of sanitation/hygiene as a health related package covering proper disposal of wastewater, garbage and animal dung through social mobilisation/education and enforcement of sanitation by-law.

Conclusion

Through experience learnt from the past two decades on sector development coupled with the limited financial resources allocated to the sector both external and internal, the Government has now realized that to continue "business as usual" would result in not only widening the gap between the served and the unserved but also gradual weakening the health of her citizens. Some specific actions have been initiated by the Government to change from the concept of "free water" to "economic good". The task ahead is enormous. The Government needs a strong political will, cooperation of her citizens and the support of the donor community to realize the change for eventual sustainable health impact and better overall economic development.

References

1. N.K. Msimbira (1989) Keynote Address 11th Annual Water Engineers Conference (AWEC), Mwanza, Tanzania
2. Ministry of Water, Energy and Minerals (1991) National Water Policy, Tanzania
3. Ministry of Water, Energy and Minerals & Ministry of Health (1991) Workshop Proceeding on Development of a Water & Sanitation Monitoring System in Tanzania, Arusha, Tanzania
4. Ministry of Health (1991) National Sanitation Guidelines, Tanzania
5. Ministry of Health (1993) Strategies for the Promotion of Environmental Health, Tanzania
6. Ministry of Health (1993) Action Plan for Attainment of Universal Sanitation Coverage, Tanzania

Strategies for effective community involvement

Pius B Mabuba

RUKWA IS ONE of 25 administrative regions of Tanzania, comprising of three districts – Sumbawanga Rural, Nkansi, and Mpanda; the regional capital Sumbawanga is administered separately as an urban council area. Some 240 of the region's 340 villages had water schemes as of December 1992, representing a population coverage of 69%. NORAD has been the major donor agency supporting the water sector in the region, ever since it funded the preparation of the Regional Water Master Plan in the late 70's. Since 1988, the water programme has been implemented largely as part of the Rukwa Integrated Development Programme (RUDEP), funded by NORAD.

Community Participation (CP) is now recognized as a key element in the water sector programme in Tanzania, being clearly stipulated in the National Water Policy (ministry of Water, 1991). In Rukwa, it was quite early realized that CP was a crucial success factor (Tschanner & Mujwahuzi, 1975). Noting the poor sustainability of schemes, it was seen that the answer was to try and make schemes as self-servicing as possible, involving the people as far as their capacity went (Raidal 1986). Since many villages could not go it alone, the idea of a partnership between the government and villages was proposed (See Figure 1).

Organization

Formal CP activities in water projects in the region started in 1981, the main objective at that time being the mobilization of village labour for schemes selected by the Water Master Plan. Over the years, CP work has been done either by engineers or technicians duly oriented to CP work guidelines, or by a special units. At the beginning, a special unit operating in the NORAD-financed Project Implementation Unit did all the GP work. It was turned into a separate 'Community Participation and Health Education Project' in 1985. The project ended in 1988, after which CP activities were coordinated by a two-person team. Since 1992, CP work has been delegated to the districts, under guidance of the Regional Community Development Officer.

Experience has shown that success of CP has depended on: perceived magnitude of the water problem by the village; timing of activities; status of the village; village leadership, and attitude of technical personnel.

Implementation strategies

Over the years, the regional water department (MAJI) has followed various strategies to improve the effectiveness of C P work; some are outlined below.

Systematic job guide

In order to ensure that there is systematic and uniform implementation, a 15-Step CP Job Guide has been used since 1984. The Job Guide covers three stages: Planning, Construction, and Operation and Maintenance. The steps relate to formal/informal meetings, site visits, discussions with village leaders, scheme attendants, etc. Specific objectives have been defined for each step, together with outcomes, and a listing of required participants. There are special forms to be filled in each step in relation to monitoring, agreements with villages, handing-over process, etc. Appropriate training is given to village leadership, members of Village Water Committees, scheme attendants, etc.

Water committees

All villages with water schemes are required to form Water Committees. This is in line with the national water policy which calls for establishment of water committees at all levels, i.e. from national, regional, district, ward, and village level. Composition of the village water committee should be at least 50% women. Some 225 out of 240 villages have formed the committee. The committee has overall management responsibility over the scheme after it has been formally handed over to the village. So far, some 35% of the village water schemes have been handed over.

Water fund

Efforts for establishing water funds in the region began in 1989, a few villages being chosen as pilot schemes. As of March 1993, some 65% of villages with water schemes have started the Water Fund. In the past, follow-up on the establishment of the funds has not been strong, but in future, this will be made compulsory. The money is to be used as villagers' contribution for planning and construction of the scheme, and, especially, in O&M. Except for a few cases, growth of the funds has been slow in most villages. Major problems facing their growth include lack of a "permanent" modality for collecting contributions, poor economic/financial base of many households, and in some cases, embezzlement of collected funds not yet banked.

District revolving stores

These have been established at District Water Engineers' offices to enable villages obtain spares for effecting repairs on their schemes. Each store is operated in conjunction with a District Water Fund. The stores were initially stocked with materials procured using RUDEP funds. Villages pay for the spares using their Water Funds. As per existing pricing policy, spares costing up to TAS 15000 will be bought at full cost by the villages; materials costing more than this amount will be subsidized by the respective District Council. The Councils or MAJI will also provide assistance with transport where necessary. From the brief experience over the one year during which the stores have been operating, the following problems have been noted:

- some District Councils have not yet contributed their share for establishment of the District Water Fund;
- Many villages have not been able to afford buying the spares, except for the very simple ones e.g. bib cocks;
- District Councils have not been able to subsidize purchases where required.

Obviously, additional measures need to be taken to ensure that the Revolving Stores are successful and that the villages buy the spares to repair the schemes.

District focus

In order to bring project activities closer to the villages and enhance beneficiaries' involvement, it has been decided as part of RUDEP policy that all grassroots operations and activities should be implemented at District Water Engineers' offices, rather than at the regional office. This is contrast to the previous practise whereby nearly all project planning and implementation was done by the regional MAJI organization. Specifically, construction of all village water schemes, and O&M issues are to be done at district level, plus all CP work. The water department at regional level will be responsible for overall supervision, plus technical support and back-stopping, and also undertake the specialized technical operations like deep-well drilling and water quality investigations.

Proper choice of technology

From the bad experience of the numerous unfunctioning motorized schemes in the region, every effort is now made to ensure that the technology employed is the most appropriate for the particular village. As per conditions in the region, the first priority is directed towards utilization of shallow ground-water by tubewells, ring wells, springs, and infiltration galleries or kanats as far as possible. Boreholes and gravity schemes are used only where studies indicate them as the most appropriate solutions. Motorized schemes are discouraged. Choice fo technology is also discussed with the beneficiaries. This use of simple technology is expected to contribute to greater involvement of the beneficiaries, and to scheme sustainability.

Involvement of women

Full involvement of women has been recognized as being crucial for effective CP. As such, deliberate steps are taken to ensure their full participation. Some of the measures taken are as follows:

- Village Water Committees are comprized of at least 50% women;

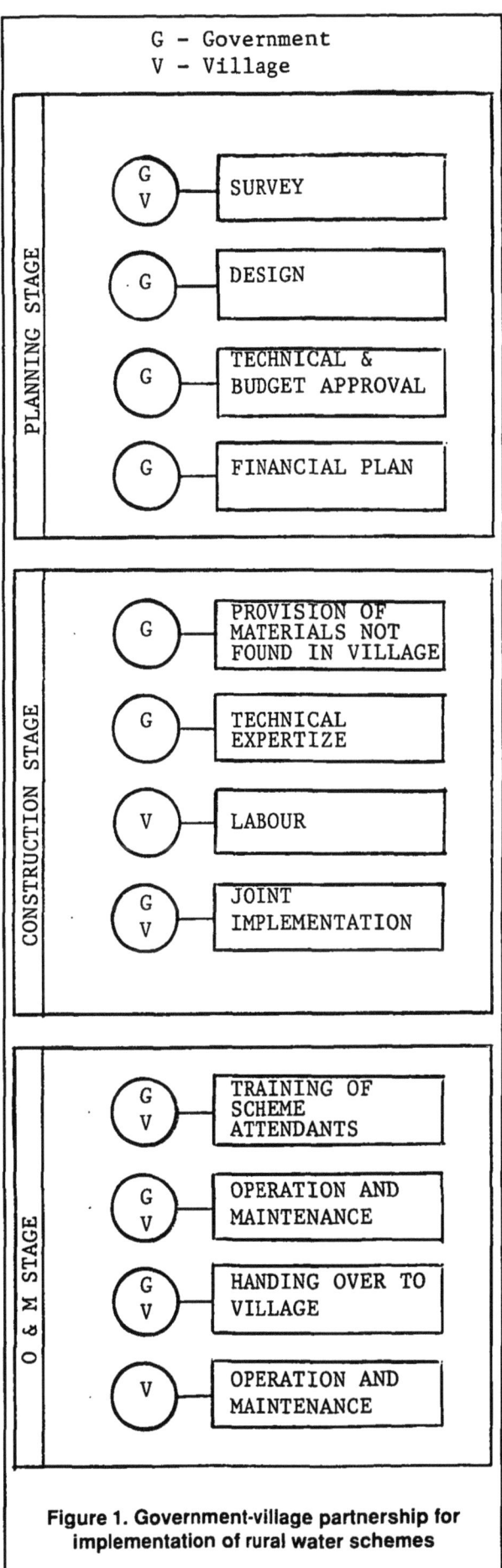

Figure 1. Government-village partnership for implementation of rural water schemes

- Separate meetings are held with women as part and parcel of the community mobilization process;
- The female members of the water committee are the ones who propose the location of water points in the village;
- CP teams in the water department at regional and district levels include several women, and where possible, are led by women.

Achievements and constraints

Successful implementation of the CP activities has shown the following achievements:

- developed sense of ownership over schemes, enhancing the spirit of self-reliance, and reducing cases of vandalism;
- reduced work-load on water department, especially with respect to day-to-day operation and maintenance of schemes;
- developed skills for self-management among beneficiaries, and greater awareness of relevant technology issues.

The following are some of the major constraints noted:

- insufficient allocation of resources to CP staff due to general budget constraints;
- pressures to achieve physical implementation targets, due to strong influence of control-oriented planning and implementation approaches (Therkiildsen, 1988);
- "lingering feeling" that water should be a free service – due to past policies.

Future direction

As far as the RUDEP Programme is concerned, CP is considered as a means and as an end. As a means, it is expected to improve the implementation of various development activities. But it is also an end in that it builds up trust and solidarity in communities, giving the people greater control over their lives and the environment. In principle, RUDEP envisages that CP shall be a non-directive programme, not fitted to a strict and suffocating time frame. This is the concept that will guide all RUDEP activities, including the water projects.

While not opposing the present situation whereby villagers are required to participate in development activities which have actually been pre-selected and pre-defined by government departments, it is envisaged that in future, villagers shall have greater say in deciding their development priorities. Every project shall in future clearly indicate the kind and level of CP aimed for, that is, whether collaborative, capacity-building, empowerment, or involvement via provision of free or paid labour.

Conclusion

The level of community participation is one of many factors affecting success and sustainability of rural water schemes. However, for the CP to be really effective, deliberate strategies are required to ensure that the objectives are attained. Planning and implementation of schemes properly incorporate the specific requirements needed for the community to participate, given the particular conditions obtaining in the village. On the other hand, beneficiaries' capacity for self-management and self-reliance has to be strengthened so that they can in turn better contribute financially or materially in the implementation of the scheme, as well as to its sustainability. Further efforts are required to ensure use of the most appropriate strategies.

References

Kauzeni, A.S., *Community Mobilization and Participation in Rural Water Development*, IRA Research Report No. 53, University of Dar es Salaam, Tanzania, 1983.

Ministry of Water, *National Water Policy*, Dar es Salaam, Tanzania, 1991.

Raidal, E., 'Some Experiences with Comminity Participation in Implementation', *in*: *Proceedings Arusha Seminar*, Ministry of Lands, Water, Housing and Urban Development, Dar es Salaam, Tanzania, 1986.

Tschanner & Mujwahuzi, *Impact of Rural Water Supply*, BRALUP Research Report No. 37, Univ. of DSM, Tanzania.

Therkildsen, O., *Watering White Elephants?* Scandinavian Inst. of African Studies, Uppsala, Sweden, 1988.

Community planning of water supplies

Ofori MacCarthy and Andrew Livingstone

THIS PAPER DESCRIBES community planning for the rehabilitation, operation and maintenance of small urban water supplies in northern Ghana, that is taking place under the Ghana Water and Sewerage Corporation (GWSC) Assistance Project.

At present, GWSC manages about 40 mechanised, piped water supplies and approximately 2,800 boreholes with hand pumps in northern Ghana. Most of the mechanised supplies were installed in the 1960s and 1970s, when the Government of Ghana embarked on a programme to expand potable water coverage to communities with populations over 1,500. Most of the boreholes with hand pumps were installed in the 1970s and 1980s, concentrating in villages in the Upper Regions.

These installations were done without involvement of the communities, and technology choices were made by GWSC engineering staff. After installation, communities were expected to pay a water tariff to GWSC, who are required by law to operate and maintain the supplies.

Many of the mechanised, piped water supplies are broken down, or shut down because of non payment of tariff. The ones still operating provide sporadic and unreliable service, and frequently do not serve large portions of the communities. The boreholes with hand pumps have had a better operational record, but most of the hand pumps now need replacement. The cost of operating and maintaining all of these supplies is a tremendous financial drain upon GWSC's meager resources. Tariff collected from community residents is a relatively insignificant source of revenue for GWSC.

In addition to water supplies installed and managed by GWSC, community residents in northern Ghana have access to many other water sources. Non-government organisations (NGOs) in particular have been active in providing shallow, hand-dug wells, boreholes with hand pumps, and surface water reservoirs throughout the area. Many residents have developed their own supplies, usually shallow, hand-dug wells and various forms of rainwater catchment.

Water supplies provided by NGOs are sometimes maintained by the NGOs at some cost to the users, but frequently it is expected that GWSC will maintain them. Water supplies developed by community residents are often inadequate in the prolonged dry season, and produce very poor quality water in the wet season. As could be expected, water-related diseases are widespread in northern Ghana.

The existing water supply development, operation and maintenance management situation in northern Ghana is not sustainable. Existing water supplies are rapidly deteriorating, and many area residents lack access to a reliable and potable water supply. The GWSC Assistance Project was formulated to establish community management of the rehabilitation, operation and maintenance of some small urban water supplies, and to assist GWSC in more effectively managing the operation and maintenance of the remaining urban supplies.

Community management strategy

The project was initially conceived as a predominantly technical undertaking to rehabilitate as many mechanised, piped water supplies as possible. Expatriate and GWSC project staff were primarily engineers and technicians, and one of the first major activities undertaken was preparation of rehabilitation designs for about 40 towns and cities. Fortunately, a comprehensive socio-economic and willingness-to-pay survey was conducted early in the project. Community residents and their opinion leaders were very vocal in expressing their displeasure with the existing GWSC water service in their communities, and their lack of involvement in GWSC's decision-making. Also, the survey revealed to project staff the extent to which people relied upon non-GWSC water supplies.

This knowledge gained by the survey resulted in a reformulation of the project activities. Significantly more resources were allocated to community development, and a dialogue was established with community and district government representatives. After some time, a community management strategy was developed that committed the project to establishing and enabling community control of rehabilitation, operation and maintenance of small urban water supplies in 14 towns.

Institutional framework

The communities, through mobilisation of Water and Sanitation Development Boards (WSDBs), have become actively involved in planning the rehabilitation and management of their water supplies. WSDBs are semi-autonomous bodies operating within the established District Assembly system legislated and enacted throughout Ghana. Each WSDB is linked to the District Assembly by a constitution, that is drafted and then ratified in each district. By-laws to enable management of the community water supply are drafted and also ratified in each district, and enacted by the WSDB and the District Administration. The District Administration also oversees the WSDB's activities in the community, provides administrative support to the WSDB, and performs a financial auditing function on the WSDB bank accounts and bookkeeping system.

GWSC's relationship to WSDBs is one of providing technical assistance and training, monitoring, and will be to provide materials and services to WSDBs on a contract basis.

Integration of WSDB activities is occurring at the district level, as they are able to link up with health committees, development committees, NGOs and other community-based organisations.

Planning for rehabilitation

Most of the communities have identified water supply as their prime need. This was clearly demonstrated at Damongo, where the chief equated the success of his reign to the acquisition of a reliable potable water supply for the town. While recognising water supply as their prime need, it is also evident that community residents are aware and concerned about the high costs associated with higher levels of service. As a direct result of this concern, low-cost technologies and less expensive levels of service were explored by the WSDBs, thus creating a blend of technologies and service levels within each community's rehabilitation plan.

WSDBs and their communities were able to assess the operation and maintenance costs and level of management required for various technology options. For example, the Damongo WSDB determined that slow sand filtration would not be feasible due to the low level of communal spirit and cooperative action in that community, whereas the Zabzugu WSDB determined the opposite for their community.

The Bole, Nandom, Jirapa and Binaba-Kusanaba WSDBs determined that solar powered water pumping was feasible and desirable in their communities, but the applications of this technology range from full solar powered pumping to conjunctive use with diesel or grid power for pumping.

Some WSDBs, such as Navrongo and Zebilla, have determined that the improvement of existing shallow hand-dug wells instead of expanding the piped system to remoter areas of the community is a more feasible plan. Most communities in the Upper Regions have opted to rehabilitate and expand the number boreholes with hand pumps to effectively serve some sections of their communities. The Yendi WSDB has included household and environmental sanitation improvements as one of their main goals in water supply rehabilitation.

Participatory design

The WSDBs have been directly involved in technical design of the water supply rehabilitations: locating public standpipes and private house connections; identifying sections where new or expanded distribution pipes were required; assisting in the calculation of institutional and commercial water demand; and facilitating topographic surveys and land acquisition for elevated storage tanks.

The location of public standpipes, private connections, hand pumps and other water demand and service points identified by the communities through their WSDBs, enabled GWSC technical teams to prepare cost estimates for operation and maintenance of the proposed supply. These costs were then evaluated within the communities to see if they were affordable, and to determine that the service levels were convenient and acceptable. In many cases, this evaluation resulted in the reduction of the number of public standpipes and private connections originally requested, and also resulted in the more widespread use of lower cost technologies.

This has helped GWSC technical teams to then prepare conceptual designs, which are then submitted to the WSDBs for their study and approval. Only after this approval are detailed designs prepared.

Financial planning

The communities are willing to pay the full cost of operating and maintaining their rehabilitated water supplies. As a tangible sign of this willingness, almost all communities have commenced collection of a deposit towards future operation and maintenance costs. A six-month deposit has been suggested. To reduce the project-borne capital costs, most WSDBs are planning to provide some volunteer labour for rehabilitation work.

WSDBs are being assisted to develop tariffs that will cover all operation and maintenance costs, and include a reserve for capital depreciation and future expansion of the water supply.

Planning for management

WSDBs have recognised the need for training to enable them to effectively plan and manage the development and delivery processes. Training needs assessments were conducted with WSDB members and for the community at large. WSDB management training is focused on financial, administrative, technical, hygiene and sanitation issues in water supply. Each WSDB member clearly sees his or her role and responsibility in managing the water development and delivery process, and is receiving training in these responsibility areas.

WSDB members are also receiving training in communications and public awareness, and are fully involved in planning and delivering public education sessions in their communities. Drama, role-play, slides, music, songs and puppets are all used by WSDB members to deliver public education.

Changes within GWSC

Formerly, GWSC technical staff planned and controlled water supplies in all small urban communities. As a result of the introduction of community planning and management, GWSC technical staff are now recognising that community members have the ability and capacity to be fully involved in water supply activities. The community WSDBs have gained confidence in their abilities, and now feel more equality in their dealings with GWSC.

Within GWSC a group of community liaison staff has been created over the past few years to help establish and then work with the WSDBs.

Conflicts were apparent between the technical and community liaison staff. The technical staff were anxious for rehabilitation to start quickly, and saw community liaison activities as causing delays in the rehabilitation. The technical staff felt they knew what type of water supply was needed in the communities, and wanted to get on with designing, procuring and installation.

The community liaison staff felt that the technical staff had an air of superiority, and were not willing to communicate with them or with community members. Technical designs being considered for the communities appeared to be far too complex and expensive for sustainable management.

These conflicts were reduced to a considerable degree by conducting interdisciplinary workshops to help each group within GWSC learn about the other groups activities, and to sensitise each group to community management objectives and methodologies. Joint monthly meetings were started, involving technical and community liaison teams, to share information and conduct work planning together on a regular basis.

Conclusion

Community planning for the rehabilitation of small urban water supplies has been proven to be effective and feasible in 12 communities to date. WSDBs are an appropriate organisation to lead the planning and management activities at the community level. Through their formal and non-formal linkages with District Assemblies, District Administrations and other community-level organisations, WSDBs are well established institutionally. GWSC has an important role to play in providing technical assistance and training to WSDBs and district governments, and by playing a role in operation and maintenance on a cost-recovery basis.

As WSDBs and district governments mature and grow stronger, planning and management of water and sanitation development at the community level will in all likelihood become the norm. This approach is probably the most sustainable arrangement for ensuring adequate water supply and sanitation facilities in small urban centres in Ghana. This approach has the potential to be expanded to rural water and sanitation planning and management through a district-based board representing residents outside of the urban areas of the district.

The spread effect of a pilot community project

Baby Mogane-Ramahotswa

GREAT ENTHUSIASM and optimism marked the inception of the Water and Sanitation Decade (1981-1990). Ambitious as it was, the decade's declaration of bringing water to all by the year 1990 was challenging, yet commendable. The fundamental issues of the time though, were a need for community participation and appropriate technology in upgrading water supply. To some agencies in South Africa, these issues were often misconstrued to mean predetermined projects for eventual implementation by communities, and cheap technology.

Without proper understanding of these concepts, it is difficult to achieve reasonable success. To many, non-governmental organizations whose geographical coverage is limited, these issues were much easier to translate into action.

Based on this premise, the Division of Water Technology (WATERTEK) CSIR, initiated a pilot project using a socio-technical approach to serve as a model for use in other regions of South Africa. A pilot project is briefly discussed in this paper followed by an in-depth discussion of a case study of a neighbouring community.

Pilot project: Kwahlophe rural ward

A pilot water supply project involving Umgeni Water and WATERTEK was initiated by Ndwedwe District Development Council towards the end of 1988. The aims of the project were to upgrade water supply and sanitation in the area, and to draw some guidelines which could be translated for use in other areas of the region.

Socio-technical feasibility study

To ensure acceptability and sustainability of the project, the team made contact with the tribal authority and other community leaders to negotiate and establish interest in a project of this nature. All leaders were unanimously in favour of the idea. Therefore, the team was given the green light to undertake a socio-technical feasibility study.

The outcome of the social study indicated a great interest in the project. Approximately 95% of the community was willing to contribute in-kind and financially to the successful implementation of the project. Technically, four distinct watersheds were identified, hence for reference purposes the area was classified into four regions viz. A, B, C and D. Three options were feasible thus:

- *Option I:* Springs protected and reticulation in all regions at R75 000 (US$23 810 - 10 May 1993)
- *Option II:* Dam, treatment, Region D and springs protected in regions A, B and C at R220 000 (US$70 000 - 10 May 1993)
- *Option III:* Borehole, chlorination, Region D, and springs protected in regions A, B and C at R190 000 (US$60 320 - 10 May 1993)

Upon being presented with the results of the technical study with the resultant options (and an in-depth discussion of their fundamental differences and cost implications), the community almost unanimously opted for the more expensive option.

The idea was that implementation would begin as soon as the community had collected R7 000 (US$2222 - 10 May 1993) based on a R30 (US$10 - 10 May 1993) contribution per family over a period of three months.

Pitfalls: community participation and management

Collection started on a very high note with people actually queuing to start donating at a community meeting. As time went on, so collection gradually waned away. In order to motivate the people, spring protection work was started. Only then did some families in arrears with their payments pay - people believe in tangibles. This trend prevailed throughout the implementation of the entire scheme.

It also became evident once the work started that only women were present to provide most of the labour needed as men were mostly away at work in towns during the week. Also, some two men nominated to undergo training on maintenance of springs, found jobs in town. Hence, a lesson learnt - women should have been elected for this chore. Furthermore, had there been some degree of paid labour, men would have remained in the ward to implement the scheme.

Because of the lack of management skills, and the low literacy level of some committee members, the management of the scheme rested entirely on the chairman, a school principal. Retrospectively, training of all committee members was undertaken. In addition, after four years of the scheme's existence, teachers and nurses emerged from the community resulting in the pooling of different perspectives in making the scheme sustainable. Also, the problem-solving ability of the committee increased.

Other pitfalls with the scheme included non-payment of recurrent costs allegedly due to some technical problems. As a result, the water minder could not receive his regular wages and no money was available to purchase diesel.

In one region some community members vandalized a protected spring twice due to family factions. It was decided to leave the vandalized springs unrepaired until the families affected resolved their disputes. With continuous back-up support by implementing agencies, the committee is able to cope with problems experienced.

It is on this point that we turn to the in-depth discussion of the KwaNyuswa case-study.

A case study: Kwanyusa

Background

Having been inspired by the neighbouring KwaHlophe water supply scheme in process, the KwaNyuswa community elected a committee and approached WATERTEK for assistance in the upgrading of their water supply. The committee had already raised R5 000 (US$1590 - May 1993) from each of the 200 families contributing R40 (US$13 - May 1993) per family.

In response to this proposal, a three member technical crew from WATERTEK visited the area to carry out a feasibility study in January 1991. In view of the proximity of this project to the pilot scheme, and the fact that the community was already motivated to initiate a project, it was deemed unnecessary to undertake a need assessment study.

In March 1991, the author held a meeting with the Water Committee to discuss the technical feasibility results. Two technical options with their financial implications were discussed so that the committee could make an informed decision. As in KwaHlophe, the committee chose the most expensive option because of its advantages. The author then assisted the committee in drawing up a proposal for funding. In fact, the committee was only given guidelines in writing a proposal which was accompanied by copies of a feasibility study report and sent to different potential sponsors for funding.

Institutional aspects/capacity building

The importance of community-based management of the scheme was emphasized and training given to the committee from the inception of the project. This committee was well established, committed and functioning at full scale.

In August 1991, a community orientation day was held at which the project was discussed in detail. The opportunity was also used to educate the community on health and sanitation issues. Of importance though, was the fact that this education was participatory in the sense that community members also presented drama and musical items on waterborne diseases. Furthermore, at this meeting one funder was able to announce a substantial allocation of money, and another handed over a cheque. The initial cost of the project was R151 000 (US$48 000 - 10 May 1993). Within a few months a major funder donated this amount which resulted in overfunding of the project. Subsequently, the project was expanded to R248 000 (US$79000 - 10 May 1993) which included individual house connections.

Labour-intensive construction

The labour-intensive approach was used in construction of key elements in the scheme - namely, the diversion weir, sand filters, reservoir, pumphouse, distribution tank and main pipelines. There was great interest when the first pipe went in. Members of the community have learned to do all the pipe laying and jointing work themselves.

Besides the temporary employment relief and other similar benefits, the pace of construction tends to be faster using a labour-based approach rather than purely voluntary labour. As a result, construction of this project took just over a year to complete. A large amount of in-kind labour has also been used for the digging and backfilling of secondary and tertiary pipelines.

Community-based management

In preparation for efficient administration of the project, the committee armed itself with paysheets, opened a cheque account upon advice of WATERTEK for convenient purchasing of goods/materials, obtained letterheads and a stamp. Except for a cheque account, all these were the committee's own initiatives. Furthermore, the committee was proactive in deciding on penalty measures for non-payers of capital costs. This emanated from the fact that some families were not interested in the beginning, and only started showing interest when the tap was first turned on.

The committee also hired and managed its labour force (skilled and unskilled) at a rate determined from time to time. Payment was strictly on attendance basis, and there was stiff competition for work. It was also decided that the treasurer and the chairman of the committee who were responsible for the day to day supervision of construction be renumerated R600 (US$190 - 10 May 1993) per month.

Construction of the scheme was completed in February 1993. A flat-rate of R8-00 (US$3 - 10 May 1993) per family to cover the recurrent costs has not been a problem. Already, the committee has a vision for the electrification of the area, in particular the pumphouse.

The spread effect

A great interest has been engendered throughout the Natal region ever since the inception of the pilot project. Communities continuously flock to the area to see and learn how the two schemes were implemented. In view of the socio-technical improvements made in implementation of the KwaNyuswa scheme, all communities visiting the area prefer similar schemes. Countless proposals from the region are pouring into WATERTEK's offices to assist communities with water projects. Also, major funders show a keen interest in WATERTEK's approach. The latter is a client to the communities, that is, communities reimburse the CSIR once they managed to raise sufficient funds. This is important in terms of community empowerment and ownership of the process.

Interestingly enough, is the fact that communities approach the CSIR being motivated, organized and committed to the project. Currently, projects of this nature are underway in the Border and Northern Transvaal regions of South Africa.

Conclusion

In concluding - most importantly, the KwaHlophe scheme has been a very good learning exercise for all parties concerned. It would appear that the future success of rural water projects will depend on a combination of community participation, appropriate technology, and training on community-based management. A close back-up support by implementing agencies, and the involvement of the local government are also essential in ensuring the smooth-running and sustainability of projects. Once the government is convinced about these essential elements, there is no doubt that most rural communities will have convenient access to safe water by the year 2000. The community's self motivation ensures commitment and sustainability. The latter was clearly evidenced by the KwaNyuswa scheme. The scheme has induced feelings of pride cooperation and self-confidence in the community. The KwaNyuswa committee, in particular the chairman - initiator of the project, talks with pride about the project - the fact which stimulates interest in other communities without schemes. Empowerment abounds in the latter scheme because mistakes made in KwaHlophe were avoided, and successes repeated.

From these definite trends it can be deducted that pilot projects could be a panacea to the development of rural water supply.

Acknowledgements

I am grateful for the financial assistance of the Director of the Division of Water Technology, Dr B van Vliet and his management team. A special vote of thanks goes to all my colleagues at the Water Care Programme for their support and encouragement.

Working with rural folk in the Northern Region

Fati Mumuni

THE PROJECT, Village Water Reservoirs, has the objective to provide potable water to the rural folk in parts of the Northern Region known as the horse-shoe area where the situation of water is very critical.

An American Psychologist, namely Abraham Maslow, noted that among other things, water and food are survival needs of humans. In many parts of the Northern Region however, water is very scarce commodity.

The Project has four sections namely; Workshop, Technical, Administration and Animation. This paper aims at discussing the activities of the Animation Section, focusing on innovations and problems. So far, the section's activities are centred on the Tolon-Kumbungu, Savelugu-Nanton and Tamale districts which have the same culture and traditions; Dagbon. All the villages are the same tribe and speak the same language. There are hardly any "strangers" in these villages.

At the beginning of the project, many of the villagers are enthused about the project but when practical project work starts mobilisation becomes a problem.

An overview of animation programme

A village's involvement with the Project starts when it applies in writing for a reservoir. Typically, the chief or the Assembly-man signs the letter. An appointment is then made for an orientation visit to be paid to the village by a combined technical/animation team which reports on the level of water need, feasibility of dam construction, alternative possible sources of water-supply and apparent capacity of the village to mobilise itself for its part as a partner to the Project.

After a village has been selected and served with its bill, the next contact is made after 50% of the bill has been paid when the social survey begins.

The social survey comprises three main parts; the water sources survey, the administration of interviews and focus group discussions.

Guinea-worm survey comes after the social survey. In the guinea-worm survey, incidence of guinea-worm is measured, knowledge of guinea-worm prevention and possession and use of filter cloths are investigated.

A period of Health Education starts with slides and discussion on guinea-worm and filtering.

Meanwhile the Technical Section busies itself with topo surveys and their findings/decisions are discussed with the village. This is the "Discussion of Technical Possibilities". Construction begins when decisions on siting and design have been agreed upon by both villagers and the Technical Section. An Animator liaises with the villagers and the Technical Section and discusses with the Chairman and "Magasia" arrangements for feeding the workers of the Technical Section.

Maintenance teams of equal numbers of men and women (4 in total) are chosen to take care of the dam. These are trained with others from other project villages with dams on health topics chosen by the villagers.

During the first rainy season, the Animation Section mobilises both the Maintenance Team Members and villagers for the Technical Section to show them how to plant and space clumps of vetiver grass on the inner and outer embankment, and how to fill erosion gullies.

Follow-up visits are made to the village maintenance teams two years after the dam construction to encourage them in sustaining their health promotion activities.

The Project has implemented a Long-term Maintenance Programme in which participating villages pay a fee for the Project Maintenance Team to visit dams twice a year for routine checks.

Below is an account of some problems the Animation faced in the execution of the above activities.

Problems

The Animation section was created in 1988 and passed through a lot of problems. The first was inability to recruit people. The section started with two animators and an expatriate, the latter as head of section. Machines were new and were working full-swing. However, construction depended on how fast the Animation section came out with social survey reports. This was a strain on the section. The method of mobilisation was through general meetings However, the villagers did not always come out in their numbers. "Participation" was at the core of project's policy and the poor response was a sign of the villagers' unwillingness to participate. Thus the team (A.S.) could travel to a village, sit in wait for the people to come out in their numbers and after sometime "walk out" of the village without working if the attendance was poor.

"Time" was a major problem. The villagers invites the team to come to their village in the morning. The project worker's morning begins at 7.30am whereas the local people's morning lasts from 6a.m to 11a.m. When we left the project at 7a.m. and got to a village at 8a.m they wondered if we slept on the way! Their morning was about to start. These were a few of the general problems of the section.

Specific problems

Orientation visit

In 1990, January, when I joined the Project I had the opportunity of assisting in that wet season's orientation visits as participant observer. Application letters came with a long list of villages written and signed by one person. This was understood to mean all the villages wanted one dam. Letters were sent to the first village in the list summoning a general meeting on a particular date. The other villages only sent representatives or never knew of these meetings. Wrong venues were taken for these meetings and the A.S. could repeat visits to these venues several times without success - the expected audience was never there. And this was disappointing. It was soon realised informally that the Assembly men sometimes wrote these applications on behalf of the villages without their knowledge. Information flow was poor and moreover some of the villages thought the venues chosen by the A.S were small villages - politically - and meetings should not be held there. Some of the villages did not have a water need - they were included in the list for economic reasons - they would contribute for the dam construction of a big village.

Another problem was, at the orientation general meeting, lots of questions were asked by the Animation team and only one man, the chief (if he was active) or the chairman, answered all the questions because when he does give a person the order to speak, he/she can not do so. The rest of the audience soon got fed up and moved out of the meeting grounds. Moreover, the very questions asked at the orientation general meeting were repeated during the social survey. The meetings were very long and tedious because our questions were too many.

In 1991, all these problems were addressed. Letters were written to individual villages even if found in the same letter giving dates for meetings. The truth came out at these separate meetings. However, where villages were close together, two separate meetings could be held on the same day.

The orientation visit questionnaire was also modified and the meetings were shortened.

Social survey

There were two questionnaires for men and women separately excluding young girls and young men. During the previous interviews it was found out that old / middle aged women did not fetch water. This was the duty of young girls (and young men in the late dry season when the girls have to travel about 8km. for water). The questionnaires were modified to reflect what the different ages knew. The landlord had questions, the women (middle aged) theirs, and the youngmen and the young girls also had theirs. The girls who were the water fetchers never wanted to answer questions while the older women were around. At focus group discussions they felt free to talk.

Initially, the social surveys were conducted over a long period of time. This was changed because the same answers were got because in the absence of animators, questions were discussed and answers were formulated to suit particular people's wishes - the opinion leaders. In November, 1991, the whole team of Animators spent three days in a village for the interviews. Nights were used for focus group discussion and better responses were obtained.

Discussion of technical possibilities

The Technical Section's proposals of technical possibilities were often not met. For instance, it may make two site proposals and the villagers would make a choice and the T.S. would come back to say the villagers' choice could not be met. Site proposals are now made after a careful study of toposurveys.

Land disputes

Because of land disputes, two villages (1990) have not had their dams even though one of them has paid fully for a dam. In 1992, in the middle of a dam construction, claim of ownership of the land led to a 2-week break in work. A decision was taken to ascertain the ownership of land as early as possible in every village. Of course, this had been done before but the Animation section always used to take the word of the village without further investigations. Now nearby villages are consulted on the ownership of the land for dam construction.

Water - hygiene education

The A.S. has to give health talks especially those related to water-borne diseases. The first question that comes in mind is how far are these talks to continue? What topics do we streamline? How long should the A.S. remain in a village after a dam has been constructed since the number of villages keep on increasing every year? The team can give talks on guinea-worm, diarrhoea and general hygiene. However, when it comes to diseases like malaria, cholera, bilharzia, skin diseases etc..., these are beyond our scope. Only the nurse animator can treat these topics.

With regard to guinea-worm discussions, the people believe that guinea-worm is in the individual's blood stream. However, there is spiritual guinea-worm cast by an enemy. Filtering is accepted because the people see living things in their water especially in the late dry season.

Feeding of the construction unit

The feeding of the construction unit by villagers has always been a problem especially when they have to work in villages in the lean season (construction takes place in the dry season when the land is firm and this may take a maximum of seven dry months). Furthermore the village youngmen who should offer free labour (one of our work policies) always feel that the construction unit are paid and should do all the work. Children are rather sent to the construction site to work instead of the young men and women. The A.S. therefore charges a small levy of $1.50 on any youngman who refuses to work (Names of the day's workers are given to the foreman by the chairman of the village). Lateness to work also fetches a small charge of 50c. a defaulter.

Dam maintenance

The major problem with the dam maintenance is the maintenance works take place in the wet season when everybody is busy on his farm. Vetiver grass transplant takes place in July immediately it rains. The villagers don't see the need to spend time on the dam wall when there is work to be done on the farm. The section therefore has to pay several trips to villages in the rainy season to encourage the people to plant vetiver grass and mend erosion gullies.

Secondly, in a dry year, it is not possible to transplant vetiver grass.

Maintenance conference

A 3-day conference of VMTs from twenty-two villages was organised by the project in August, 1992 where topics related to dam maintenance, environmental influence of the dam, and the economic use of the dam were discussed.

Conclusion

Animation is an uphill task. To try to effect change in people is not easy. Yet, there are interesting times in the work. The beginning of the work in the village is always interesting:- That is when the team enjoys maximum co-operation. However, after dam construction, mobilization becomes difficult - It is always the chief, chairman, "magasia" and VTMs who attend to animation needs. For the rest of the community once they get water, they have finished with us. There is always "the other side of the coin" - And our joy has been the gratitude often expressed.

Acronyms

A.S	Animation Section
T.S	Technical Section
VMTs	Village Maintenance Teams.

Community participation: Umgeni Water's approach

Adrian Wilson

THE WIDESPREAD FAILURE of development projects in Africa in the 1960's and 70's led to a re-assessment of the approach needed and of the critical success factors for implementation of such projects. A number of reasons were postulated for this phenomenon, perhaps the most important one being that the communities which these projects should have been serving had not been sufficiently involved in the planning, implementation and administration phases. This led to a lack of ownership and hence ultimately, sustainability of the projects. This perception has certainly been felt in South Africa although the trend has if anything been slower to reach these shores than for other countries.

This change of emphasis had undoubtedly resulted in a paradigm shift for those involved in development projects. It necessitates a new approach which many agencies and actors have been struggling to come to grips with. One of the reasons for this is that the skills required to facilitate effective community participation are quite different from those historically associated with the implementation of projects. Nevertheless, the principle of community involvement is now widely accepted but it is more the practical aspects of how to most effectively achieve this that people are wresting with. In many ways the old autocratic approach of implementing projects is more intuitively appealing to people with a hard technical background, even though the long term results of this are often disastrous. It is fair to say that "technocrats" are often not great communicators and thus they struggle with many of the concepts and practicalities of community participation.

Most of the material outlined in this paper is as a result of the experiences that Umgeni Water has had in the planning, implementation and administration of schemes in rural, peri-urban and informal settlement areas. This paper thus unashamedly has an emphasis towards water and sanitation schemes. Nevertheless, it is believed that a lot of the principles and philosophies embodied in this paper have generic application to other development projects. Umgeni Water has now had approximately 12 years experience in this work which has been particularly concentrated in the last five years. This is as a result of the implementation of Umgeni Water's Rural Areas Water and Sanitation Plan (RAWSP).

The emphasis of this paper is intended to be primarily towards the practical aspects of how to facilitate community participation and this is reflected in the main body of the report.

Why community participation?

The primary objective for community participation has already been mentioned as being that of sustainability of projects. This has an obviously very pragmatic motivation from any implementing agency's point of view. However, within this overall framework and objective are a series of other reasons and motivations for adopting this sort of approach. Ten reasons advanced for community participation by White (1981:11) are as follows:

1. With participation, more will be accomplished.
2. With participation, services can be provided more cheaply.
3. Participation has an intrinsic value for participants.
4. Participation is a catalyst for further development.
5. Participation encourages a sense of responsibility.
6. Participation guarantees that a felt need is involved.
7. Participation ensures things are done in the right way.
8. Participation uses valuable indigenous knowledge.
9. Participation frees people from dependence on other skills.
10. Participation makes people more conscious of the causes of their poverty and what they can do about it.

The above list clearly indicates that one is trying to address the issues of building ownership in the community, utilising local resources, and developing appropriate and hence efficient solutions.

In the South African context there are also additional motivations for community participation in that one is looking (as is common with many other countries) to transfer skills to the community and to develop institutional capacity. There is also a need to build goodwill with communities which have often experienced decades of poor treatment and frustration. The lack of a democratic process in South Africa also emphasises the need for properly conducted community participation which will enable some sort of democratic input to occur in the development process. Lastly, one must say that in the current mood prevailing in the country the community participation option is ultimately politically much more acceptable and more likely to facilitate reconciliation.

How does one achieve community participation?

Groundwork

The experience in Umgeni Water has been that one needs to do a lot of homework on and with communities before one can proceed to the implementation stage. This involves a process of getting to know the community and of the dynamics that operate within it. In this regard it is particularly important to identify potential stakeholders and power groups that operate within a community. This is made more difficult by the fact that one cannot always take for granted a claim by an individual that he or she represents the community.

Community liaison

One has to set up some from of channel of communication between the agency and the community. An important technique for achieving this is by the medium of public meetings. These can be frustrating and difficult forums but are nevertheless essential in terms of ensuring some sort of reasonably democratic contact with the community. They are not a practical means for detailed community participation but are nevertheless extremely useful for feedback to the community when key points in a project are reached and very important decisions need to be made.

For the detailed process of consultation and involvement in decision making, one has to work through some sort of community structure. If there is such a structure in place within the community then this can be used but it is important initially to carry out some discreet enquiries to try and establish the credibility, and support, of such structures. The structure involved could take the form of a Water Committee, Development Committee, Tribal Authority, Civic, Residents Association or indeed any other committee that can be seen to have support in the community. If no suitable structure exists then it is necessary to try and encourage the community to elect a committee to serve as the liaison body between the agency and the community. Again, the most appropriate mechanism for facilitating this is the public meeting.

Regular meetings will need to be held with the community committee throughout the project but the frequency will vary depending on the project phase and needs.

Throughout all the above phases the committee should continually be acting as a channel of communication between the community and the agency. This is clearly ideally a two way communication process though committees tend to have to be reminded of this, particularly with regard to their duty towards the community side.

Agency representatives

There is a need to devote a great deal of time and patience to the community liaison process. It is one of those things that is very difficult to rush but time invested in the early stages will be worthwhile in the final analysis. One therefore needs to have people available who can spend a significant amount of time in the community and who can communicate effectively. One needs to have people involved who have the right sort of outlook and attitude for this sort of work, not everybody is disposed or "cut out" for it.

The ideal background for third world development projects is a mixture of technical and social skills. This is an unusual combination and thus it is sometimes necessary to make use of teams of people who have the different skills in combination working closely together.

Key skills for facilitating community participation were identified in the RAWSP (Institute of Natural Resource (INR), 1991: 60-61) as follows:

Communication skills
Educational skills
Development skills
Facilitation skills
Evaluation skills

The pursuit of empowerment

If one is to achieve empowerment within the community then often the agency has to be involved in a process of skills transfer and institution building throughout the community participation process. The latter in particular is a severe constraint in many communities due to the political legacy of the past. Of course it goes without saying that the community have to wish to be part of the process if it is to be successful, after all, one of the ultimate indicators of empowerment is the ability of the community to make choices even if these are sometimes not particularly liked by the agency with which they are interacting.

It should also be emphasized that the building of capacity within the community is a two way process whereby both parties have certain well defined responsibilities and tasks to undertake. Empowerment cannot be achieved by the community sitting back and letting the agency do everything for them. Problem solving is an important part of the learning process (Flanagan, 1988:16-17). One should also be aware that a lot of the participation process involves negotiation with the community and it is submitted that this is a very healthy situation. It is apparent that a situation where the community just blindly agrees to everything often results in an end product which is ultimately not sustainable.

Some practical suggestions

Attitudes

When trying to establish a relationship between the community and the agency, which is essential in the process of community participation, the attitude of both parties are key variables of which one needs to be aware. As mentioned previously one must accept the fact that the agency's efforts are often viewed initially with mistrust and hostility and one has to have patience at the early stages in order to be able to work through this. Community attitudes in this regard can certainly be changed over time but one needs to develop a climate of trust and establish a track record before this can be achieved. Only then can a healthy relationship and partnership develop.

Agency representatives must therefore be prepared to invest a lot of time and must have an empathising approach and genuine interest in the welfare of the community. In the words of Van Wijk-Sijbesma (1989: 12) "more time is needed to decide things with others than for others". Another important aspect to mention is the concept of who is the client. It would appear obvious that the community must be the ultimate client yet some bodies and agencies appear to become confused about this issue, even though it is fundamental to the concept of providing a service and ensuring the recipient is happy with the end product.

Political

There are exceptions and sometimes hidden agendas but generally speaking a need for development is accepted by most communities and political groupings. The more tricky issue however is the question of the divisions within the communities which is particularly common in the Natal situation. There is no easy solution to this problem and it must be recognised that a development project has the potential for both unifying and splitting any community. In an ideal world one would hope that a development project could bridge gaps and bring people together and Umgeni Water has had some successes in this regard.

In other cases however, one sometimes has to adopt a strategy of "divide and rule" in order to be able to achieve any success in a particular area. This would involve setting up separate committees representing different areas within the overall development area. Discussions and negotiations can then occur with the various groupings on an individual basis. It must be stated that this is not ideal and certainly inefficient but is sometimes the only way to make progress. A critical success factor in this process is to maintain an apolitical stance as far as possible. You must have the freedom to be able to talk to all parties at all times.

In many areas, particularly the more rural, traditional structures are in place and these should be involved at an early stage in the participation process as they often retain considerable authority in the community. It is interesting to note this is a factor in many parts of Africa (White, 1981: 134-5). Traditional structures vary a lot in their management style, some adopt a hands off policy once they have given a project their initial blessing, others like to be heavily involved throughout. The former is often preferable since it appears to improve the speed of decision making.

Communication

Communication with any community is difficult and something at which one has to work very hard. Because of the problems with illiteracy in underdeveloped communities one has to recognise that the spoken word counts for a tremendous amount. In this regard public meetings are particularly useful but also the radio is a very important means of communication in developing communities, although it can be expensive. The youth can play an important role as the literacy among the young people tends to be much higher than amongst the older people. Umgeni Water have found that communication via the schools appears to be a very effective means of getting things back to the parents and the community as a whole.

Another thing that one has to accept in view of the fact that key decision makers are absent during the week is that a lot of crucial work has to occur over weekends both in the case of public meetings and also committee meetings. Public meetings are one of those "necessary evils" of community participation. They can be extremely difficult to manage but are nevertheless an essential tool. A good rule for public meetings is "expect the unexpected". As a result, cool heads are needed for agency representatives, who must be able to think quickly and make decisions where appropriate.

Negotiation

It has been mentioned that throughout the whole participation process one is ideally looking at developing capacity within the community. In this regard one has to be realistic and realise that there are many issues which are negotiated throughout the process of the project. This means that the normal principles of negotiation apply such as knowing what is negotiable and what is non-negotiable, recognising differences, whilst emphasising commonalities etc. etc. Another principle of negotiation is that one should not underestimate or overestimate the capacity of those parties one is negotiating with. We have found that there are often some very shrewd and able negotiators active within communities.

Another important principle of negotiation is to recognise that every community is different and a flexible approach is thus required (see for example Shandu and Wilson, 1992). Solutions that work in some areas are not successful in others. An innovative and lateral approach is needed in some cases as "canned" solutions may not be appropriate.

Opportunities

In spite of the multitude of problems that present themselves in working with communities there are also many opportunities which can be capitalised upon. A good example of this is the low employment in many areas and usually there is tremendous excitement and demand for any employment opportunities. The principle of employment of local labour wherever possible, and as a minimum for the unskilled work, should thus be widely accepted by development agencies. This creates the potential to develop tremendous goodwill amongst the community whilst also adding considerably to the potential for ownership of the project. There is often a surprising amount of local skills that are potentially available within communities. One should look at opportunities for promoting entrepreneurship within the community by strategies such as labour only contracts and labour based construction. There are also often ongoing roles in administration and maintenance that can be very effectively "picked up" by local community members after the projects have been implemented. This ensures continuity and again facilitates ownership.

The liaison structure that one works through in the community participation process has the potential to become a tremendous ally in the development of the project and also as a catalyst for further development in the area. The water committee of today could become the local authority of tomorrow. To promote this one has to try and look at means whereby one can emphasise the status of the committee wherever possible. This could involve things such as encouraging other development agencies to work through the committee and also ensuring that all important decisions go through the committee. One does also however need to almost set the committee certain tests and check on these to try and determine whether the committee is doing its job and playing the correct role.

Conclusion

To Umgeni Water, community participation has become an accepted and natural part of project planning, design and implementation, maintenance and administration. We believe that it is such a useful process with so many potential benefits and spin-offs that it almost becomes an end in itself. Some of these that immediately spring to mind are the potential to develop good will, empowerment and real improvement in the quality of people's lives. All these benefits can accrue to any agency or body adopting the community participation and partnership approach. It is not however an easy panacea and considerable time and effort must be devoted to achieve success. A real commitment is needed by the development agency and its representatives as there will be many setbacks and frustrations along the way. This paper has attempted to give some practical suggestions to assist those interested in following the community participation approach. In the final analysis, it is the only route to follow if South Africa's vast development challenges and targets are to be achieved.

References

Flanagan, *Human Resources Development in Water and Sanitation Programmes*, The Hague. IRC, International Water and Sanitation Centre, 1988.

Institute of Natural Resources, University of Natal, *Rural Areas Water and Sanitation Plan*, Volume 1, Pietermaritzburg, Umgeni Water, 1991.

Shandu and Wilson, *Systems used to Administer Water Supply Schemes in Rural and Informal Settlement Areas*, Pietermaritzburg, Umgeni Water, 1992.

van Wijk – Sijbesma, *What Price Water? User Participation in paying for community-based water supply*. The Hague. IRC, International Water and Sanitation Centre, 1989.

White, *Community Participation in Water and Sanitation*. The Hague. International Reference Centre for Community Water Supply and Sanitation, 1981.

SECTION 2

GROUNDWATER

Siting of sanitary landfill and faecal treatment

Robert R Bannerman

LONG-TERM PLANNING for improvement in solid waste and night soil disposal of Accra, the capital city of Ghana, called for assessment of the suitability of different locations within the Greater Accra Metropolitan Area and the adjoining Ga District, to serve as sanitary landfill and faecal treatment plant sites.

Methodology adopted for the study comprised the search for technical information and its review; interviews with Government authorities, chiefs, stool elders, land owners and users; and reconnaissance surveys.

The proposed sites were identified through interpretation of aerial photographs; and the study of large scale geological and topographical maps. As much as possible field observations and measurements were made to establish the ground truth. The sites were then plotted on large scale maps, marking the coordinates of prominent edge points.

The survey

Details of the survey included the description of the following:

- location and accessibility of proposed sites, not only in terms of distance from the generation points but also taking into consideration the class and interconnection of road and rail network.
- topographical characteristics: indicating the elevations of natural valleys (lowlands), their trends and gradients; and any modifications of these that have resulted from man-made activities.
- the lithology and structure of soil and overburden materials and their thickness; and structural features of the bedrock or any outcrops; and also of any made-up ground.
- geohydrologic conditions: such as estimation of the permeability of the different soil and rock types; and the presence of permanent or ephemeral streams, rivers and ponds and their discharges; and ascertain whether the sites are prone to flooding.
- groundwater occurrence and its utilization and measurement of groundwater levels and flow directions.
- current and future land use.

Other aspects of the survey included determination of:

- distances to nearby developed communities.
- the ownership of the said lands.
- the size of the areas (in hectares).
- prospective fill volumes (in cubic meters) for the landfill.
- As some of the sites are near the Accra International Airport, to assess potential restrictions by air traffic regulations.
- For faecal treatment sites, assess among others the volume of water in streams and lagoons to take in the effluent.

Assessment

Landfill

Accessibility to all 4 sites studied are generally adequate; as they are approachable by tarred and second class laterite roads. But they fall within increasing distances, varying between 7 to 16km from the waste generation centres.

Topographical characteristics

All the proposed sites are in valleys with gentle slopes which are closed on one side, with the other side opening into adjoining larger valleys.

In 3 cases ridges mark the rim of the valleys making them "contained".

The bottoms of these valleys have been extensively modified through excavations for sand and gravel; and this activity has more or less created additional fill volume.

The sites have topographical characteristics which allow rainwater and surface water to drain off; hence have "high" suitability for land-filling purposes.

Geohydrologic conditions

All the sites are underlain by Precambrian metamorphic rocks comprising gneisses, schists, quartzites and quartz schists.

The bedrock is overlain by shallow weathered and decomposed materials consisting mainly of sands and sandy clays, and in some cases, transported alluvial deposits. The thickness of the overburden varies between 0 and 30m, averaging 10m over most of the areas.

Special regard was given to the permeability of the sub-soil as this determines the amount of water which can trickle through these layers and potentially contaminate the groundwater.

Generally geohydrological conditions are favourable as none of the areas has good groundwater sources and potential that could be impaired.

Surface water discharge

3 of the valleys are occupied by ephemeral streams and are prone to flash flooding. In one case, there is a perenial pond at the lower end of the valley.

Distance to nearby developed communities

Only one proposed site is within 800m from a human settlement; the rest, located more than 1000m away from are regarded as averagely suitable.

Ownership of land

Three of the four sites are owned by Chiefs and Stool Elders. This would not make land use negotiation easy. However for landfill operations, even over a longer period, it does not seem to be necessary to own the land. A contract only for the utilization of the land will make negotiations easier. In the case of the one site owned by Government negotiations could be much easier.

Official land use plan

The kinds of land use which have been officially allocated to the area by the Government Planning Authorities were reviewed. Also considered was the time distance from now till when these plans should be realised. 3 sites were assessed as of "high suitability" as there are no specific plans for immediate residential or industrial use.

Distance from airport/flight path

The distance of the proposed sites from the airport and the flight path was a crucial argument for assessment. 3 of the sites were found to be near aircrafts flightpath, thus birds, like vultures could become a serious safety problem for flight operations.

Area fill volume

This concerns the life span and the initial investment costs to start a new landfill. A large site with options of a deep fill or a high build up of refuse are highly desirable. An average suitability was given to 3 sites, of anticipated life span of 5 years.

In addition the possibility of future expansion was assessed. 2 sites which offer such opportunity were assessed as having "high" suitability; and 2 other sites with rather unknown future land use plans were shown as "average". One site with very limited future expansion plans showed "low" suitability with regard to this factor.

Availability of suitable cover material within the vicinity

Fill-and-cover should be a standard for the operation of a sanitary landfill site. Daily covering removes strench, minimizes fly nuisance and discourages birds on the landfill.

Easy availability of cover materials like sand, gravel, laterite rubble or even saw dust, is highly appreciated for landfill operations. At 3 sites such materials are easily available within the vicinity and hence these sites were assessed as of "high" suitability.

Assessment

Faecal treatment sites

3 proposed sites were assessed, with regards to their location, accessibility, prevailing wind direction and especially, nearness to a large volume of water in a stream, lagoon or pond to ensure dilution and drainage away of the effluent.

All the sites are accessible by tarred road. 2 of them, considered suitable, are located on hill slopes with sufficient gradient to allow flow of effluent into a large perennial volume and flow of water.

Airflow both in the valleys and on the hill sides is quite adequate for removal of odour. The sites are not near any developed neighbourhood and there is no new development planned in their immediate vicinity.

Conclusion

The attributes of each of the 4 proposed landfill and 3 faecal treatment plants sites were compared and contrasted. This exercise was based on criteria and ratings developed from arguments, factors and issues under the different conditions.

Grades of assessment were thus established, following rules for assessing qualitatively measurable characteristics. The assessment result of (low, average or high) suitability was then linked to a numeric system.

In this manner, all the prospective sites were screened and sites that met most of the criteria and thus most appreciated were chosen.

The result of the study highlighting the principled arguments, would be presented to the Waste Management Division of Accra Metropolitan Authorities to help them decide on the final choice and approval of the sites.

Nitrate pollution of groundwater sources at Oyarifa

N K Sekpey and S A Larmie

AN INTEGRATED COMMUNITY development programme comprising sufficient and safe water supply and sanitary facilities are necessary for ensuring an improvement in health conditions and living standards in rural communities.

Groundwater is the most cost-effective source of water supply in the rural areas and is exploited through the use of hand dug wells or boreholes depending on the depth at which it occurs. Though this source if properly developed and utilized, is generally safe for human consumption, natural or artificial contaminants in the water could seriously hamper its use. The main contaminants of concern include high salinity, excess iron and high nitrate levels. Whereas the first two are essentially naturally induced, nitrate contamination in groundwater is invariably caused by man's interference with the environment.

Oyarifa is a village in the Greater Accra region with a population of about 2000 persons (refer to Fig.1). The village had no reliable source of water supply and the situation was particularly critical during the dry seasons.

Two ponds were constructed to store storm runoff for domestic use but due to their poor quality the villagers are constrained to walk a distance of about 2km to fetch water from an abandoned quarry. Also, a hand dug well constructed under a UNICEF/GWSC assisted programme had not helped to alleviate the water problems facing the community because of poor water quality.

In consequence, the Water Resources Research Institute was contracted to provide a borehole fitted with a handpump for the Community in July 1992. The borehole was drilled to a depth of 64m in phyllite and yielded 5 litres per minute during air-lift (refer to Fig.1); the static water level was 4.8m.

Chemical analysis

Samples collected from the borehole source were analysed for chemical quality. Although most of the chemical parameters were within WHO limits the results revealed a serious nitrate-nitrogen problem in the order of 80 mg/l which is far in excess of the 10mg/l guideline value given by the World Health Organisation (WHO,1984) for potable purposes. Nitrate in water is of health significance since it is toxic when present in excessive amounts and it has been reported to cause methaemoglobinaemia in bottle-fed infants in some cases. For older age groups this problem does not arise but there is a possibility that certain forms of cancer might be associated with high nitrate concentrations.

Initial inferences were that the nitrate pollution was due to two unlined pit latrines, one located 200m south of the borehole and the other 150m to the north. Each pit latrine is located adjacent to one of the ponds in the village. In order to investigate the extent of the nitrate problem water samples were again collected from the borehole, hand dug well, two ponds and the quarry for chemical analysis. The results of chemical analysis for the 5 samples are given in Table 1.

Although the surface water sources showed low nitrate level, high levels were shown by the borehole and hand dug well. A preliminary explanation was that there was a single source of nitrate pollution, possibly the pit latrine adjacent to Pond 1 which was affecting the two sources. The intense plant activity observed through the excessive green colouration in the ponds suggest high intake of nutrients by the plants; leading to the low levels of nitrate in the surface water sources.

Geologic model and groundwater occurrence

The area falls within the Dahomeyan-Togo contact zone with granitic gneisses overlying phyllite or foliated schist. The overburden consists of lateritic clay and sandy clay. Existing borehole records reveal groundwater occurrence at the weathered rock and bedrock interface and in fractures in the bedrock.

The 64m deep borehole at Oyarifa intercepted a water zone at a depth of 53-54m in phyllite (refer to Fig.2). The borehole log shows the lithologic sequence; the interface between weathered rock and gneiss as well as the interface between gneiss and phyllite were dry.

The hand dug well is fed by a shallow unconfined overburden aquifer localised around Pond 1. The overburden around the borehole was however dry. Site reconnaissance also revealed a topographic high between the borehole and hand dug well; this would provide a groundwater divide for at least the shallow unconfined flow system, ensuring that the flow around the hand dug well was in a direction away from the borehole. Considering the location of Pond 1 in an upper slope area underlain by shallow gneiss bedrock, there should have been good yielding fracture aquifers in the bedrock assuming it was sufficiently tectonised. This was not the case however.

Furthermore, the existence of hydraulic communication between the shallow unconfined groundwater flow system and the confined flow system in phyllite would have to be via fractures in the gneiss, ensuring that the gneiss-phyllite interface yielded some water to the borehole. This would have confirmed preliminary explanation of the nitrate pollution problem.

Since the gneiss-phyllite interface was dry, possibly indicating that there is no hydraulic communication as

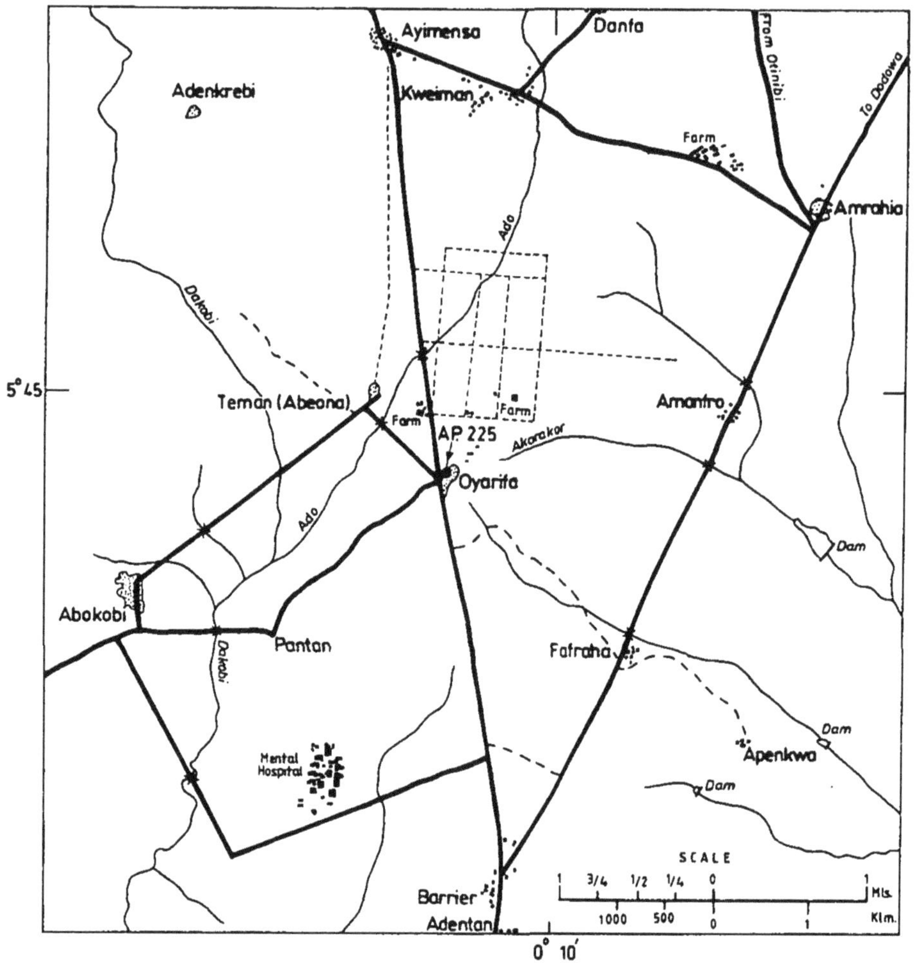

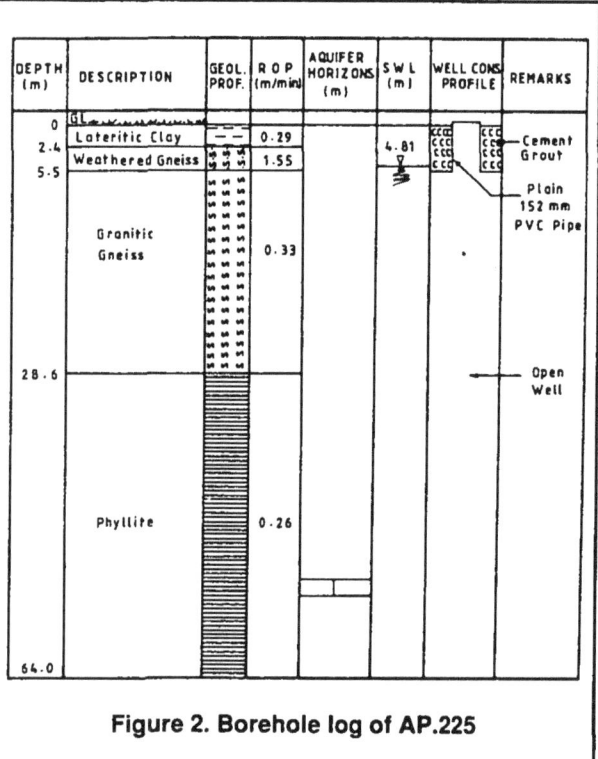

Figure 1. Location map of Oyarifa - showing position of borehole AP.225

Figure 2. Borehole log of AP.225

postulated above, it is suggested that the source of nitrate pollution in the borehole is different from that in the hand dug well. This implies a dual nitrate problem comprising a point source of pollution from the pit latrine in the water-bearing overburden and a larger scale diffuse source from an agricultural source. This postulation is partially borne out by the reported use of chemical fertilizer in a nearby farm complex in an area which is underlain mainly by phyllite (Fig.1).

In contrast, chemical analyses of samples obtained from several boreholes in the granitic gneiss area gave acceptably low nitrate levels.

Conclusion

Results of chemical analysis of groundwater samples obtained from a new borehole and an existing hand dug well have revealed a serious nitrate pollution problem at Oyarifa.

The initial indications were that the nitrate pollution was due to a pit latrine. However, an evaluation of the geologic structure showed that the two groundwater sources were fed by separate flow systems.

It has therefore been suggested that there is a dual nitrate problem comprising a point source from the pit latrine and a diffuse agricultural source.

Despite the problem, controlled use of the borehole source for adults is possible provided a continuous monitoring of nitrate levels is carried out. The best option for improving the quality of the hand dug well source is to relocate the pit latrine.

The experience at Oyarifa confirms the need for a comprehensive investigation of such problems in the rural environment to ensure that large-scale problem sources are not masked.

**Table 1.
Nitrate levels in water sources**

Source	Nitrate-nitrogen (mg/l)
Borehole	84.6
Hand dug well	118.5
Pond 1	1.7
Pond 2	< 0.01
Quarry	1.5

References

1. Hiscock K.M., Lloyd J.W., Lerner D.N. and Carey M.A., 'An Engineering Solution to the Nitrate problem of a Borehole at Swaffham', Norfolk, U.K, *Journal of Hydrology*, 107, pp. 207-281, Elsevier, 1989.

2. Soil Survey and Land Research Centre, Cranfield, Groundwater Vulnerability, Map 6 Mansfield, Severn-Trent Water, 1987.

3. Water Resources Research Institute, *Report on Groundwater Exploration at Oyarifa village*, Ghana Organic Agricultural project, July 1992.

4. World Health Organisation, *Guidelines for Drinking-water Quality*, Vol.1, Recommendations, WHO, Geneva, 1984.

Low-cost GIS for water resources

Richard M Teeuw

WITH PERSONAL COMPUTERS (pcs) becoming both more powerful and less expensive (£0.5K-£1.5K), many organisations dealing with water resources can now use low-cost Geographical Information Systems (GIS) to improve their performance. A GIS is a computer-based package for merging, analysing and modelling data that can be displayed as maps. This can provide a rapid and powerful means of examining the many sets of data held by agencies concerned with water resources.

For a given district, datasets as disparate as population density, types of aquifer, water quality and well locations can be merged and presented as computer-generated maps, summary tables or graphs.

A good GIS will incorporate powerful analytical tools, covering general statistics, spatial statistics, image analysis and time series analysis. This allows for modelling and the examination of 'what if' scenarios for proposed developments, producing a wealth of new information for decision makers.

'Top of the range' GIS packages, such as ARC-INFO, are very expensive, falling in the £10K to £20K price range: they also require highly trained staff and costly computers (usually workstations in the £5K to £10K range). That said, the analytical power of such systems can be very useful at central government level, allowing the storage, merging and analysis of many diverse national databases.

However, at the organisational level of a water resources department, the 'top of the range' GIS may be a waste of money, given the amount of data available and the information required.

The Applied Geology Unit at the University of Hertfordshire, England, has found two low-cost GIS packages to be particularly useful for water resource management in developing countries:

- AEGIS, based on points, lines and areas linked to a database (a 'vector' GIS) costs about £150 and is very 'user-friendly', being Windows-driven. Microsoft Windows costs about £60, but is often included free of charge with new pcs. Rather than buying a database package such as dBASE IV, costing about £300, the Notepad facility of Windows can be used to complile and manipulate databases. An example of an AEGIS application, accessing the Windows/Notepad database to illustrate variations in river corridor width, is given below.
- IDRISI, a grid-based (or 'raster') GIS developed with UN support, costs about £180. It is menu-driven and relatively 'user- friendly'. Furthermore, IDRISI has a powerful image processing module, allowing the use of digital satellite images, or scanned-in aerial photographs, for mapping. A version running on Windows, due in 1994, should facilitate the swopping of datasets between IDRISI and AEGIS: that will be a major saving, as most integrated raster-vector GIS packages cost at least £4K.

Both of these GIS packages come with comprehensive and relatively easy to use manuals. Both have tutorial packages that guide new users through key GIS usages. Anybody who has been able to master a word-processor or a data base package, should be able to use IDRISI or AEGIS after only two or three days of work on the tutorials. This is an important point, often neglected: what use is a GIS - no matter how multi-functional - if nobody knows how to use it?

Horror stories abound, regarding organisations purchasing very powerful GIS packages to do relatively simple operations. Perhaps these purchasers quite reasonably assume that:

- if a GIS package is very expensive, it must be very good: a powerful and and easy to use GIS that can readily tackle a wide range of problems, might be envisaged. Here a 'Buyers Beware !' warning is needed: as a rule of thumb, the more things a GIS can do, the more complex it will be: its ease of use decreases accordingly. This was particularly true of ARC-INFO, until the recent release of ARC-VIEW, a more user-friendly Windows-based version.
- if so many other major organisations have purchased a given 'top of the range' GIS, it must be good. This view is true for some GIS packages, particularly those developed in the past five years with integrated raster-vector formats. It is not so true for 'top of the range' packages that were first developed in the 1970's: these rely heavily on their historical leadership of the GIS market and on extensive marketing to maintain their dominance. Most of the low-cost GIS packages are primarily for educational purposes: making a profit is not a major consideration. For instance, the IDRISI GIS is being developed by Clark University in the USA with support from the UN Institute for Training and Research and the UN Environment Programme. Producing an effective, and affordable GIS for developing countries is the main objective of the IDRISI project. It may be that insufficient emphasis has been given to marketing low-cost GIS packages, to make them as well known as highly advertised 'top of the range' packages. ARC- INFO is often heralded as the most widely used (vector) GIS in the world: IDRISI's status as the most widely used raster GIS in the world, gets hardly any mention.

To illustrate the value of low-cost GIS in water resource management, some examples of GIS applications are outlined below:

- Input of digital satellite images (raster format), allowing the production of, (a) regional maps of roads, rivers, settlements, land use and vegetation; (b) interpretive maps of soils, regolith, rock types, lineaments and landforms. The costs of satellite data range from £0.02 to £0.50 per km^2.
- Detailed mapping using aerial photography or other thematic maps (relief, geology, hydrogeology, geophysics, etc): either as vector input using a digitising board and electronic pen, costing at least £300; or input in raster format using a scanner, costing at least £150.
- Storage and analysis of tabular data, for instance: meteorological data (rainfall, evapotranspiration); hydro-geological data (water supplies: surface, subsurface; borehole data: water table depth, water quality, recharge rate); socio-economic data (population census; land/water holdings; local sources of income). Rapid conversion of tabular data to more readily understandable map data.

The end result of using a GIS with these datasets would be a series of maps that summarise a few key aspects of the entire dataset that might be of interest to the user. These thematic maps are produced by the GIS overlaying one set of data over another and 'sieving' out unwanted or irrelevant data. Shown below are two data sets, or layers (lineaments and regolith cover) that have been merged using IDRISI. The resulting map highlights the prefered locations for new water wells: sites where intersecting lineaments occur in areas of deep regolith. An addition GIS analysis could involve:

- interogating a field survey database to find which local villages have inadequate water supplies;
- automatically producing a new map that only shows those new well sites that are within 5 km of villages with inadequate water supplies - thus targeting the wells that should be started first.

Up until now, the usage of GIS was somewhat elitist, often restricted to a very expensive system, with a few GIS experts, without whom the system would be virtually useless. The availability of both low-cost GIS, plus low-cost computers, makes GIS technology available to a much wider range of staff than before.

The accessibility and integrity of water resouces data should also improve using a low-cost GIS on many pc's with many users; as opposed to a 'top of the range' GIS where data is processed by a few office-based experts (often with no relevant field sampling experience). With a low-cost GIS, data can be both collected and processed by the same member of staff, enabling he or she to check for errors: this is very useful, as it is difficult and time-consuming to track corrupted data back to its source. The same members of staff can also run 'what if' scenarios for their sampling areas: this leads to a better interpretation of the GIS results because of their first-hand knowledge of the sites.

Low-cost GIS packages have been available now for the last ten years or so. Powerful low-cost computers now make these information systems affordable in developing countries - where they will be of particular value to the water resources sector, with its many disparate data sources. Major improvements in efficiency result from using a GIS to store and analyse large volumes of data, produce new maps and up-date old maps. Decision makers should also benefit from using a GIS: they can rapidly examine new combinations of data, as well as selecting only the most relevant data from an often bewildering array of tables and maps.

It has only been possible to present a very brief outline of the potential benefits of using low-cost GIS packages in the water resources sector; for further information, an FAO publication dealing with GIS and aquaculture (Meaden and Kapetsky, 1991) is very useful. Other relevant publications are those of the Centre for Earth Science Studies, Kerela, India (1991), detailing the use of GIS in low-cost surveys of land and water usage; and Asabere (1992), detailing the use of IDRISI for environmental assessments in Ghana.

References

Asabere, R.K., 1992, 'Environmental assessment in the mining industry using GIS', *Mapping Awareness*, Vol.6, No.10, 41-44. (ISSN 0945-7126).

Centre for Earth Science Studies (CESS), 1991, 'Panchayat level resource mapping'. 38pp. From: *Co-ordinator, PRM Programme*, CESS, Po.B.7250, Thuruvikkal P.O., Trivandrum-695 031, Kerela, India.

Meaden, G.J. & Kapetsky, J.M. 1991 'Geographical Information Systems and remote sensing in inland fisheries and aquaculture', *FAO Fisheries Technical Paper: 318*, (Rome), 262pp. (ISBN 92-5-103052-9).

Supplier's addresses: IDRISI, Graduate School of Geography, Clark University, Worcester, MA 01610, USA. Fax: USA 508 793 8842. AEGIS, AU Enterprises Ltd, 126 Great North Road, Hatfield, Herts., AL9 5JZ, UK. Fax: UK 707 273684.

SECTION 3

HEALTH AND DISEASE

Community-based surveillance in GWEP, Ghana

Dr Sam Bugri

GUINEAWORM DISEASE *(Dracunculiasis)* infects man through ingestion of contaminated water. It is therefore a good indicator of the non-availability of safe water supply to rural dwellers in endemic areas. Not much was known of the extent of the guineaworm problem in Ghana until a sample survey was conducted in the Northern Region of the country in 1987 and a National Case search in 1989. (Ghana GWEP, 1989) Both surveys demonstrated that passive surveillance was recording less than 5% of cases occurring.

A national effort towards the eradication of dracunculiasis (guineaworm disease) began in 1988 with assistance from Global 2000 Inc.Cater Presidential Centre. A plan of action (POA) was made which set 1993 as the target date for eliminating transmission of *Dracunculiasis* in the country. This paper describes briefly our experience in improving surveillance of guineaworm disease in Ghana.

Passive surveillance 1960-1986

As early as the 17th century, European explorers reported seeing cases of guineaworm disease along the coast of what is now Ghana.

Studies by various medical scientists, beginning with Dr B.B. Waddy in 1956 (Waddy, 1956), established that the disease was widely distributed throughout the country and that it had a significant adverse impact on agricultural productivity. (Belcher et al, 1975).

Guineaworm disease began to be included in the routine monthly reports from peripheral health units throughout the country to the Ministry of Health in 1960. Most of these cases reported by health institutions were individuals seeking medical care because of secondary infection, cellulitis resulting from the worm braking within the tissues and arthritis. The absence of effective treatment for the disease certainly did not encourage patients to seek specific treatment at health care facilities.

In 1981 the author estimated that <30% of the cases actually occurring in the Northern Region of Ghana were reported through the health facilities (Bugri, 1981). More recent information suggests that in fact less than 5% were being reported. Up to 1986, these routine reports averaged a total of about 4, 000 cases annually for the entire country. (Fig.1) These early data did confirm that guineaworm disease was occurring throughout Ghana, but gave only a vague indication of geographic priorities among the ten regions, and even that sense of priorities turned out to be significantly flawed, when compared with the results of the national case search conducted later.

Active surveillance (1987-1992)

In 1987, a sample survey of guineaworm disease was conducted in the Northern Region of the Country. The results of this survey were largely responsible for the increase in officially reported cases that year to 18, 398 (Figure 1). In 1988, all Regional Directors of Health Services were instructed to mount active control and surveillance measures for the disease. A sample survey of villages was also conducted in Eastern Region (WHO, 1989). These efforts did not cover all endemic villages, nor did they record all cases. Following the 2nd African Conference on Dracunculiasis in Accra March 1988, extensive publicity and education through the news media and by health workers resulted in increased reports of "outbreaks" of guineaworm by villagers. Health workers supported by Global 2000 GWEP visited many such villages in response to these reports, to count cases, give some first aid, and begin educating the population about the disease. Thus the number of officially reported cases in 1988 rose to over 70, 000.

Guineaworm disease is basically a disease of the rural dweller invariably living in small settlements far away from health facilities.

For the purpose of eradication a more sensitive system of surveillance had to be developed. (Richards, Hopkins, 1989). Community/village based data collection was thought to be the best source of information. In 1989, we began to train district and zonal guineaworm programme coordinators (each district is divided into zones consisting of a number of villages), who in turn identified village volunteers (VV), one in each endemic village, to report cases of guineaworm disease from their villages monthly.

The training covered the life cycle of the disease, preventive measures and the education messages to be propagated. They were taught how to fill the surveillance forms and if they were illiterate they got a school boy to do the writing of the names. The village volunteers who are the key people in this programme were carefully selected by the community. He/she had to be self employed and a respected member of the community, such that they would accept educational advice from him/her. The District Coordinators are health workers but the zonal and village level workers are volunteers. 6, 515 endemic villages were identified during the national Case Search.

The surveillance system is outlined in the flow chart.

The national case search

At the time of writing the Plan of Action the actual endemicity and geographic distribution of cases was not known. A National Case Search was therefore conducted at the end of 1989 to ascertain the true extent of the problem.

The National Case Search was a massive mobilization exercise, involving over 12000 people and lasting 60 working days. Field investigators were selected at district and zonal level for their knowledge of the area and the language, and trained specifically for the case search. A total of 19, 759 villages or 92% of all villages in Ghana were visited, and a questionnaire administered to a reliable resident(s) such as the village chief, head teacher, pastor, chairman of the local village development committee, etc. The screening questions asked on these visits were whether the respondent(s) knew what guineaworm was, and whether there had been any cases of the disease there within the past year. A village was classified as endemic if one or more cases of guineaworm was seen in that village within the past year and if the village was using surface water for drinking. If a village was classified as endemic a suitable volunteer was identified and coached on how to use a booklet provided by the programme to do house to house interviews to record the names of all inhabitants who had guineaworm disease during the past year. These volunteers were given one or two days to complete the search, depending on the size and spread of the village, and they were monetarily rewarded for their work. Respondents were also asked whether blood in urine was common in the village - a rough indicator of urinary schistosomiasis.

This national case search, which cost approximately US$ 50, 000 (provided to the Ministry of Health by the USAID mission to Ghana and Global 2000), revealed the location and incidence of guineaworm disease in Ghana for the first time (Fig.2). We were no longer looking at the tip of the iceberg and imagining what was below the surface. We now had the whole mountain before us, and what huge mountain it was.

Monthly surveillance: 1991,92,93,

Since 1991 the progress of the programme has been monitored by the monthly surveillance results. The reporting rate has improved from an average of 67% in 1991 to over 94% by the end of 1992. Some regions are receiving reports from 100% of all known endemic villages. (Fig.3)

The monthly reporting system allows us to monitor progress month by month comparing figures with the same period the year before.

We are able to see the effects of stepped up interventions of the previous year reflected in the following year. The monthly surveillance also gives a clear picture of the seasonality of the disease. (Fig.4)

The National pattern is greatly influenced or dominated by the figures from the Northern Region which contributes >50% of cases.

The monthly surveillance coverage does not only provide incidence of cases it is also an indicator of the amount of education reaching the affected areas. Each time the VV visits a house to collect information on cases he also gives education on preventive measures and inspects and replaces their cloth filters.

Discussion

The main challenge in the programme now as far as surveillance is concerned is to make the village based monthly reporting achieve 100% coverage every month and to change from monthly reporting to daily reporting or at least within 48 hours. In the case containment mode all cases must be managed individually and on time.

- If a case is detected before the worm emerges, the worm must be surgically extracted - the know how and the ability is available in all health facilities in endemic areas.
- If a case is detected after the worm has emerged, controlled emersion and occlusive bandaging is advised.
- Vector control using Temephos is routine but if a case is reported from a village, it can be assumed to have infected cyclopods in the ponds of the village with guinea worm larvae which would mature to infective stage in 10 days. Temephose is applied to the ponds to knock out the infected cyclopoids.

Currently plans are being made to make full use of the experienced village based workers to improve surveillance on other indicators and improve broader Primary Health Care services in these villages.

The effectiveness of this programme in Ghana has benefited greatly from the awareness and enthusiasm generated by the head of state, Fl-Lt J.J.Rawlings.

As a sign of his personal interest and support, he spent 8 days in June 1988 visiting 21 endemic villages in six districts of the Northern Region, teaching villagers about the disease and demonstrating how to filter drinking water through a clean cloth. The two visits to Ghana by former U.S President Jimmy Carter have also helped to focus attention on the disease.

We do not see a substitute for complete surveillance in guineaworm eradication programme. Already in January this year we could see from the monthly returns that we are likely to see cases in 1994. If we had relied on annual case count/search, this will not be revealed until 1994.

References

1. Belcher, D.W., Wurapa, F.K., Ward W.B., Lourie, I.M., 'Guineaworm in southern Ghana: its epidemiology and impact on agricultural productivity'. Am J Trop Med Hyg 24:243-249 1975.
2. Bugri S.Z., *Guineaworm: an indicator of the quality and quantity of rural water supply in northern Ghana*. (Dissertation) The Ross Institute, London School of Tropical Medicine and Hygiene, 1981.
3. Ghana GWEP National Case Search Summary 1989.
4. Richards F.O., Hopkins, D.R., 'Surveillance: the foundation for control and elimination of dracunculiasis' in Africa. Int J Epid 18: 934-943, 1989.
5. Waddy B.B., 'Organization and work of the Gold Coast medical field units'. Tran R Soc Trop Med Hyg 50:313-336, 1956.
6. World Health Organization, *Dracunculiasis: Ghana*. Wkly Epidem Rec 64:16-19, 1989.
7. Richards F.O., Hopkins D.R., 'Surveillance: the foundation for control and elimination of dracunculiasis in Africa'. Int J Epid 18: 934-943, 1989.

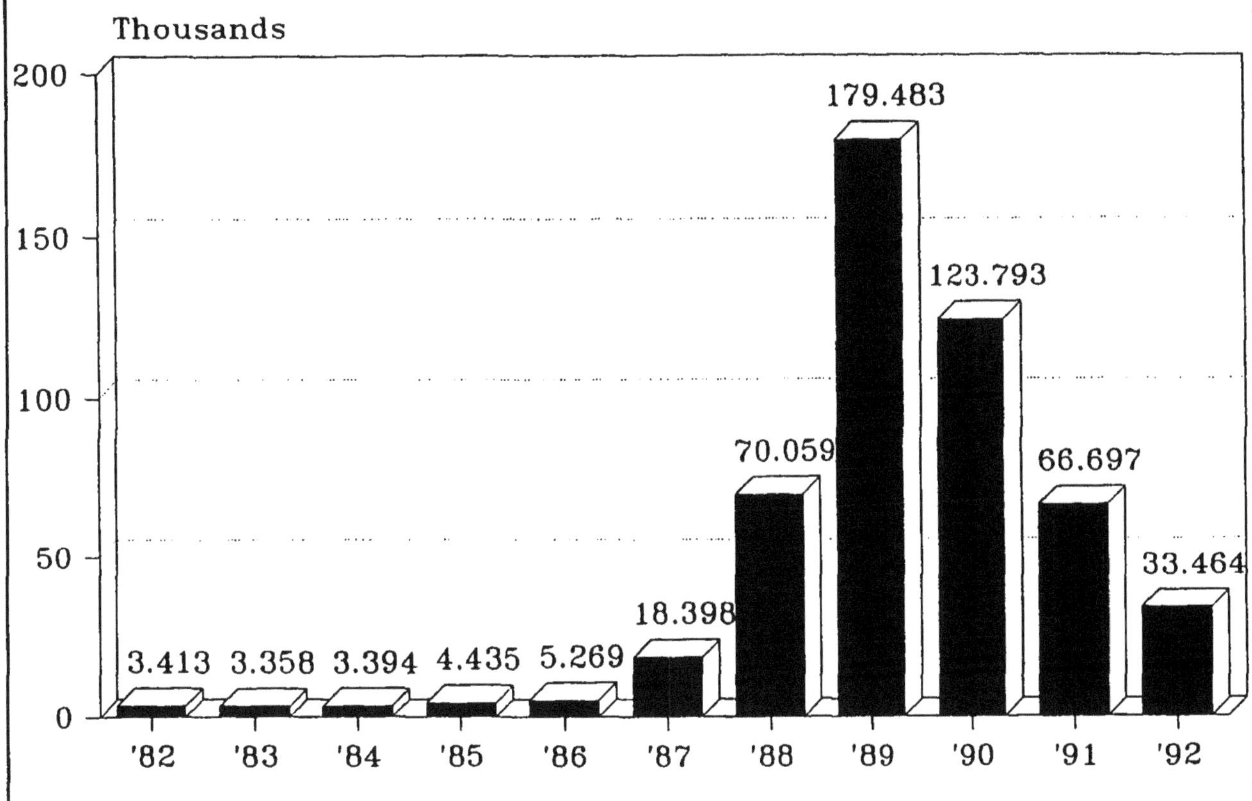

HEALTH AND DISEASE: BUGRI

Figure 1. G.G.W.E.P. number of reported cases by year

Year	Cases
'82	3,413
'83	3,358
'84	3,394
'85	4,435
'86	5,269
'87	18,398
'88	70,059
'89	179,483
'90	123,793
'91	66,697
'92	33,464

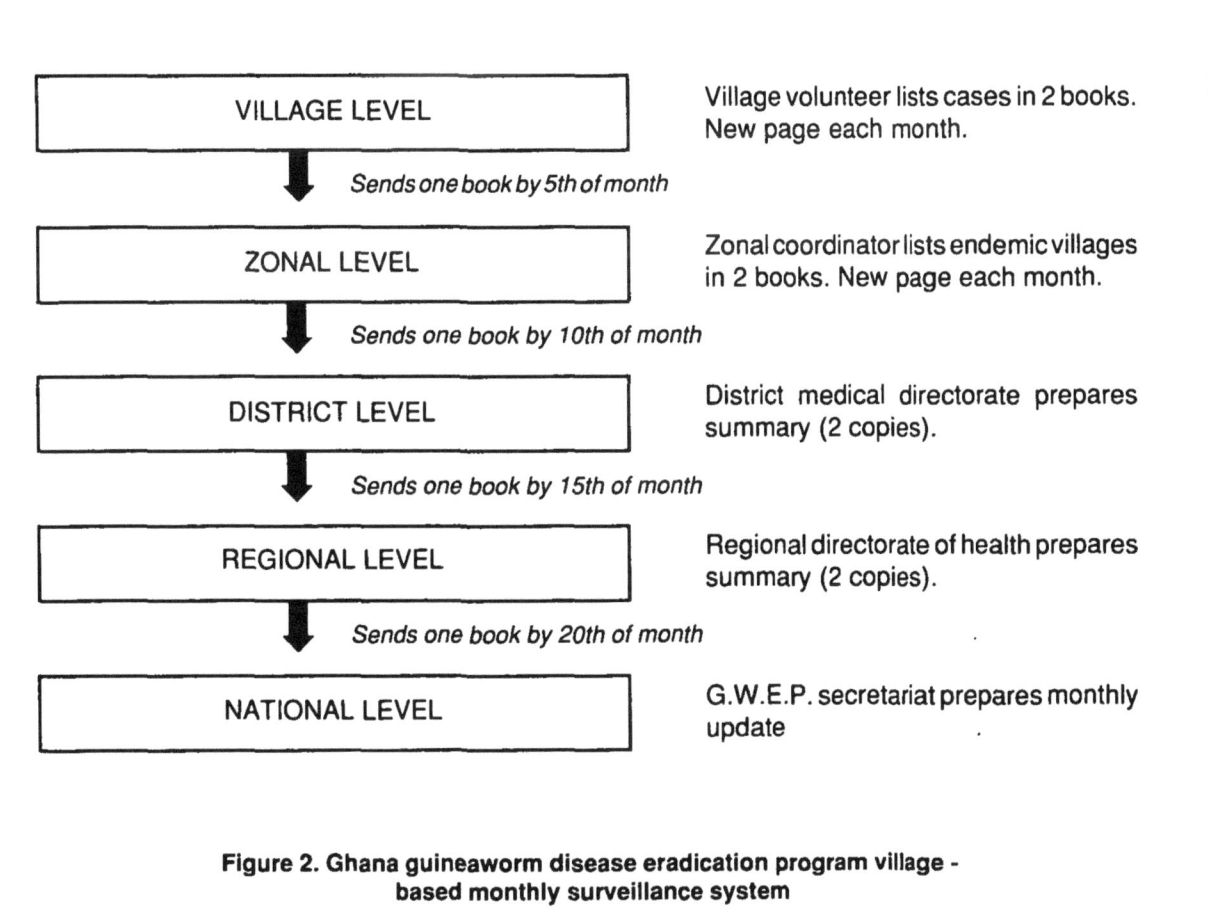

Figure 2. Ghana guineaworm disease eradication program village - based monthly surveillance system

- **VILLAGE LEVEL** — Village volunteer lists cases in 2 books. New page each month.
  - *Sends one book by 5th of month*
- **ZONAL LEVEL** — Zonal coordinator lists endemic villages in 2 books. New page each month.
  - *Sends one book by 10th of month*
- **DISTRICT LEVEL** — District medical directorate prepares summary (2 copies).
  - *Sends one book by 15th of month*
- **REGIONAL LEVEL** — Regional directorate of health prepares summary (2 copies).
  - *Sends one book by 20th of month*
- **NATIONAL LEVEL** — G.W.E.P. secretariat prepares monthly update

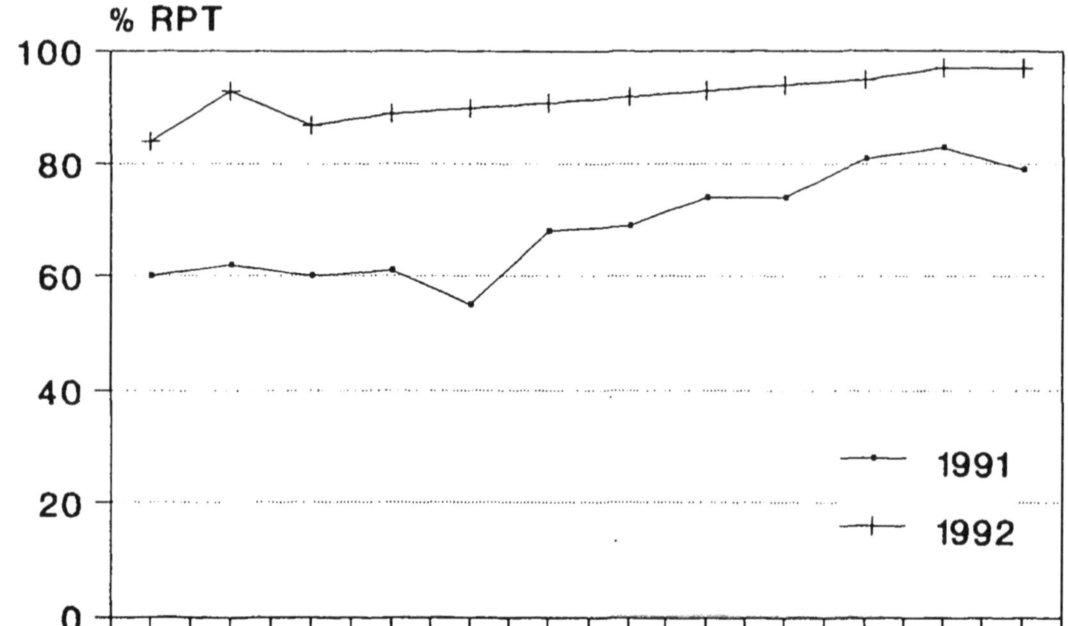

	JAN	FEB	MAR	APR	MAY	JUN	JUL	AUG	SEP	OCT	NOV	DEC
1991	60	62	60	61	55	68	69	74	74	81	83	79
1992	84	93	87	89	90	91	92	93	94	95	97	97

Figure 3. G.W.E.P. % of end. vill. rpt. monthly 91 & 92

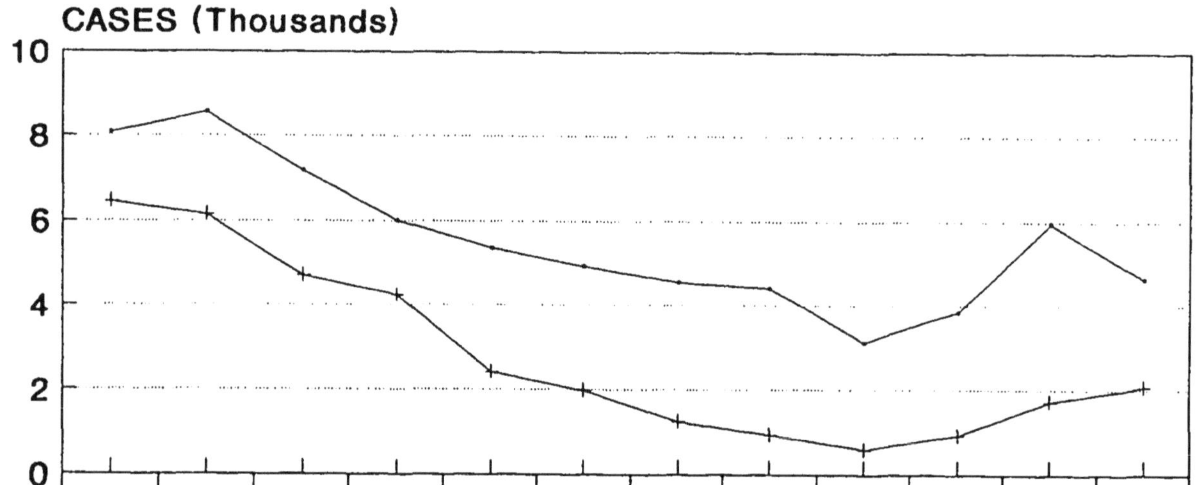

	JAN	FEB	MAR	APR	MAY	JUN	JUL	AUG	SEP	OCT	NOV	DEC
1991	8.071	8.546	7.16	6.007	5.378	4.949	4.572	4.428	3.124	3.85	5.959	4.653
1992	6.446	6.138	4.727	4.246	2.423	1.985	1.258	0.936	0.593	0.926	1.719	2.066

Figure 4. G.W.E.P. 1991 & 92 reported cases

Guineaworm eradication – is the target attainable?

Dr Sandy Cairncross

The challenge of eradication

It is not often that humanity decides to eradicate a disease. It has only happened twice before; we failed with malaria, but succeeded with smallpox. If we do not achieve it with guinea worm, we shall only have ourselves to blame. Water engineers should be delighted at the choice of a disease in whose eradication they can play an important role, and which we can hardly fail to vanquish in the end.

Why can Guinea worm be eradicated, when malaria could not? The disease is different from malaria in several ways. The parasite does not live beyond its one-year transmission cycle, so that there are no long-term carriers. The disease is easily diagnosed, and cases are easily prevented from passing it on to others. Moreover, it does not have a vector which can fly about, transmitting infection over distances.

However, the most compelling evidence that Guinea worm is eradicable is that it has already disappeared, with or without human assistance, from a number of countries. In addition to Central Asia and most of the Middle East, these include at least two African countries, Gambia and Guinea, where the last indigenous cases were seen some 20-30 years ago. In these countries, there has been no health education aimed specifically at the disease, so that the credit for its elimination is doubtless due mainly to improvements in drinking water supply.

Even in those countries where the disease is still endemic, water supply has no doubt had an important impact. Here in Ghana for example, it has been suggested that the relatively low incidence of Guinea worm in Upper Region is associated with the large number of boreholes drilled there in the last two decades. However, such associations need to be interpreted with caution. Sanmatenga, in neighbouring Burkina Faso, is the province with the most boreholes, but also the most endemic for Guinea worm.

Why building water supplies is not always enough

There are three potential reasons why water supply improvements are not always sufficient on their own to eradicate Guinea worm disease. *The first* of these reasons is that the water supplies themselves may not continue to function, especially if arrangements are not made to ensure their maintenance.

A cautionary illustration of this is provided by the Ivory Coast, which in the 1970s was in the vanguard of African countries providing rural water supplies. Between 1973 and 1985, 12,500 new boreholes were installed in rural areas, at considerable expense. As a result, the annual number of Guinea worm cases reported fell from 67,123 in 1966 to only 1,889 in 1985 (Anon 1987). In 1991, however, the national case search found an estimated 12,690 cases (République du Côte d'Ivoire 1992). The search methods may not have been exactly comparable, but another factor behind this recrudescence of the disease became clear last year, when UNICEF carried out a survey of handpumps in three of the most endemic subprefectures of the country. They found that over half of them were out of order, in spite of a recently-concluded handpump rehabilitation project in the area.

The second reason is that the water supplies, although functioning, may not be used by most of the affected population. The most common cause of such non-use is that the new source of water is not conveniently situated, close enough to people's homes. In many of the sahelian settings where Guinea worm is endemic, such as the Mossi Plateau of Burkina Faso, the settlement pattern is so dispersed that a "village" is not a cluster of households but an administratively-defined area of land including many scattered homesteads. With Burkina's mean population density of 33 per km^2, an average "administrative village" of 1000 people covers an area of 30 km^2. In such circumstances, to speak of a village being "covered" by a single borehole, or even by two or three, is clearly absurd.

The allocation of water supplies to villages by government rural water supply programmes often begs the question of defining a village. Even when the settlement pattern is relatively nucleated, an official village often includes a number of outlying hamlets, which may or may not be permanently inhabited; they may only be occupied during peak periods of agricultural activity. Unfortunately, these periods, principally the rainy season, often coincide with the peak season for transmission of guinea worm disease.

The provision of water supplies to small hamlets is often hard to justify, for two reasons. First, it is expensive in terms of cost per head. A borehole costing $10,000 dollars for only 80 people is not a very cost-effective option, especially if they will not be there to use it all year round. Second, the maintenance of water supplies in small and often transient communities is even harder to sustain than in larger and more permanent ones; they are less likely to be able to have the funds, the skills or the degree of organisation to carry out repairs successfully.

On the other hand, the prevalence of Guinea worm disease is often significantly higher in the smaller villages. In one study of the disease in Northern Zou Province, Bénin, it was found that villages of 100 or fewer inhabitants typically had 3 to 4 times the rate of infection of the largest villages, with populations of 800 or more (Tempalski 1991).

This may explain why water supply programmes which are targeted to areas endemic for guinea worm often fail to reach the specific villages which are most affected. In the Bénin project for which the above data were collected, villages with a population below 150 (and thus likely to have a high incidence of guinea worm) were specifically excluded from the borehole programme (Yellott 1990).

The third reason why water supplies on their own are not always sufficient to eliminate dracunculiasis is that they are not always used *exclusively*. A great deal of guinea worm infection is acquired by casual use of unprotected sources when away from the home, particularly when working in the fields. This is borne out by the common findings, particularly in communities where the incidence is relatively low, that adults, particularly farmers, are more commonly infected than children (Belcher 1975; Cairncross & Tayeh 1988), and that people who travel away from their village of residence are also at greater risk of infection (Tayeh 1992).

In these circumstances, provision of a water supply only prevents the infection which is acquired at home. It tends to turn a high-incidence village (where, say, 20% or more of the population are infected) to a low-incidence village (typically 10% or less), but not to eliminate the disease.

What kind of water supply?

It has long been known that ponds are ideal sources of guinea worm transmission (Onabamiro 1952, Scott 1960). Transmission is particularly intense in the last weeks before a pond dries up. The cyclopoid population is increasingly concentrated in an ever-decreasing volume of water, and the infected cyclopoids, which tend to sink to the bottom (Onabamiro 1954), are increasingly likely to be scooped up as the village pond becomes a shallow puddle.

Compared with the endemic areas of India, where large numbers of stepwells facilitate transmission, the number of ponds per head of population in Africa may seem very small, suggesting that chemical treatment of the water to kill the cyclopoids might be a viable strategy, as it has been in India and Pakistan. However, it presents a number of practical problems.

First, the peak season for guinea worm transmission is in the rainy season, when access to endemic villages and their ponds is difficult. Treatment would have to be applied monthly at least to be effective; probably once a week.

Second, it is much harder to calculate the volume of an irregular pond than of a rectangular Indian stepwell, although this is essential in order to estimate the dose of insecticide required. Moreover, the pond volume varies with time; rainfall after the treatment could dilute the insecticide to harmless levels.

Third, the high suspended solids content of an African pond diminishes the efficacy of an emulsified insecticide such as temephos, which tends to be adsorbed onto the suspended particles. Chemical treatment of ponds under African conditions, even by highly qualified research teams, has been found to be of questionable effectiveness (Guiguemdé *et al.* 1990, Sullivan 1991).

What is less often documented is that many, if not most of these ponds are *man-made*. Hafirs, rehabilitated in Sudan for drinking water supply, and small dams, built in Ghana for agriculture or livestock watering, have been implicated as important sources of the disease (Tayeh and Cairncross 1991). However, the traditional man-made ponds are far more common. A similar technique is found across much of the African endemic belt; a depression is dug in the path of a small ephemeral watercourse, and the excavated earth is piled on the downstream side to make an improvised dam. Often, clay is used to make an impermeable lining to the reservoir. The result, a sort of dew pond, is known as a "dugout" here in Ghana, a "boullie" in Burkina Faso, and an "atapara" in northern Uganda.

Man-made ponds seem to be more important in transmission than natural ponds. Steib and Mayer (1986) working in Burkina Faso, found that only 9% of guinea worm patients had drunk from large natural ponds and 5% from small natural ponds, but that 85% had obtained their drinking water from small man-made ponds in the previous year. Such small man-made ponds are often dug near fields. They do not last throughout the dry season. Since the fields remain in the same place only for a limited number of years, new ones are continually being dug. Steib and Mayer (1986) found as many as ten per village, although the number may be less in other parts of the region.

If water engineers were willing to develop low-cost methods of improving these sources, such as infiltration wells or trenches, or simple sand filters such as those developed in Bangladesh and West Bengal (DPHE 1989, AIIHPH 1993), they could help not only to reduce the incidence of Guinea worm, but to prevent it entirely.

However, the rural water agencies in many endemic countries have been reluctant to consider technologies other than boreholes for application on a wide scale. Indeed, they have often planned on the basis of quite inadequate numbers of boreholes, particularly when they are to be equipped with hand pumps. A recent borehole project aimed at guinea worm control in a highly-endemic area of Nigeria, for example, provided less than one handpump on average for over 1000 people. Even the best handpump is unlikely to last very long under such heavy use - unless some of the potential users are put off by the queue and decide to collect pond water instead!

The other strategy

The strategy which has appeared recently to be producing the quickest results in a number of countries has little to do with water supply improvements. It is to set up a network of village workers who carry out monthly surveillance of cases, and also offer health education to their neighbours on prevention of disease. There is a good scientific basis for linking surveillance with health education, because recent research has shown that those suffering from Guinea worm in one year are at least five times more likely than others to have it again a year later

(Tayeh et al. 1993). The cases detected by surveillance are therefore the ideal targets for health education.

The most obvious component of the health education is the distribution of cloth filters to remove cyclopoids from water. However, there is circumstantial evidence to suggest that the dramatic reductions in cases recently found in Ghana, Nigeria and elsewhere have resulted not primarily from the use of filters (the reductions have often started to occur *before* most of the filters were distributed), but from other changes in behaviour, possibly relating to the contamination of water sources by those who have the disease.

Whatever the reason for their success, these networks of village surveillance workers are an immensely powerful tool. In Ghana, more than 90% of the village volunteers are sending in monthly reports on time, and they are willing to collect information on subjects other than Guinea worm. It would be sad indeed if this system were to be dismantled when the Guinea worm is eliminated, and not be extended to support other health initiatives, such as the eradication of polio.

Geographic information systems

One question raised by the masses of data collected by village-based surveillance is the lack of a good database on the villages in the endemic countries, or on their locations. The question is complicated by the problems of defining a village, mentioned above, and by the way villages move and change their names. However, computer-based geographic information systems (GIS) make such evolving datasets easier to handle (Clarke *et al.* 1991).

Once the data on village locations have been entered into a computer using a GIS, other information from case searches or from monthly surveillance returns can easily be added. From the database, it is easy to print maps at any scale, showing the areas needing priority targeting for disease control, the high-incidence villages lacking water supplies, the locations of health centres and schools which can be used to support health education, and so on.

Such maps have applications which can go far beyond the health sector. Made available to implementers at local level, they can be extremely valuable aids to programming. At national level, they are an effective means to convince decision-makers. The challenge is to create the sustainable capacity to maintain and use these systems, hitherto largely dominated by expatriates, and to generate demand at local level for the maps they can produce.

The water sector has much to offer here. In most of the francophone endemic countries, the water ministries have already compiled geographic databases of the locations of all villages, with and without boreholes, often using the satellite-based global positioning system to verify the precise locations of villages where maps are unreliable. I imagine that there have been similar developments in the English-speaking countries. Water ministries could provide a great service to other sectors if they helped to disseminate such data and the skills needed to manage them.

Which target?

Whether or not Guinea worm will be eradicated by 1995, it is clear that it will not be with us for many more years; the number of cases has fallen substantially in each of the last few years, as the other contributors to this session can tell you.

I would suggest, however, that we should be aiming for a more ambitious target, which is to use the eradication of Guinea worm as a means to establish sustainable systems which will serve us long after the Guinea worm has gone. These would include handpump maintenance systems, systems for providing water to small villages, including the low-cost upgrading of existing water sources, systems to monitor whether water supplies are used, systems for village-based public health surveillance and the one-to-one delivery of health information, and systems for the management and analysis of information about individual villages at national level.

Whether the endemic countries will reach *that* target is a more challenging question to answer. To a great extent, it is up to all of us.

References

AIIHPH (1993) Proceedings of the workshop on sanitary protection and upgradation of traditional surface water sources for domestic consumption. Calcutta: All India Institute of Hygiene and Public Health, and UNICEF.

Anon (1987) Dracunculiasis: Ivory Coast. *Weekly Epidemiological Record*, **62**, 23, 169-170.

Belcher D.W., Wurapa F.K., Ward W.B. and Lourie I.M. (1975) Guinea worm in southern Ghana; its epidemiology and impact on agricultural productivity. *American Journal of Tropical Medicine & Hygiene*, **21**, 1, 243-249.

Cairncross S. and Tayeh A.T. (1988) Guinea worm and water supply in Kordofan, Sudan. *Journal of the Institution of Water and Environmental Management*, **2**, 3, 268-274.

Clarke K.C., Osleeb J.P., Sherry J.M., Meert J.-P. and Larsson R.W. (1991) The use of remote sensing and geographic information systems in UNICEF's dracunculiasis (Guinea worm) eradication effort. *Preventive Veterinary Medicine*, **11**, 229-235.

DPHE (1989) A report on the development of a pond sand filter. Dhaka: Department of Public Health Engineering, Government of Bangladesh, and UNICEF.

Guiguemdé T.R., Gbary A.R. and Ouédraogo J.B. (1990) Lutte contre la dracunculose: problematique du traitement chimique des points d'eau au Temephos (ABATE) en Afrique. *Publications Médicales Africaines*, **110**.

Onabamiro S.D. (1952) The geographical distribution and clinical features of *Dracunculus medinensis* in southwest Nigeria. *West African Medical Journal*, **1**, 159-165.

Onabamiro S.D. (1954) The diurnal migration of cyclops infected with larvae of *Dracunculus medinensis* (Linnaeus), with some observations on the development of larval worms. *West African Medical Journal*, 3, 189-194.

République du Côte d'Ivoire (1992) *Situation Epidémiologique du Ver de Guinée en Côte d'Ivoire*. Abidjan: Ministère de la Santé et de la Protection Sociale.

Scott D. (1960) An epidemiological note on guinea-worm infection in north-west Ashanti, Ghana. *Annals of Tropical Medicine and Parasitology*, 54, 32-43.

Sullivan J.J. (1991) Field evaluation of slow-release temephos formulations for control of copepods in dracunculiasis elimination programs: update September 1991. Atlanta, USA: Centers for Disease Control.

Tayeh A.T. (1992) The Epidemiology of Dracunculiasis in Parts of Sudan and Ghana. PhD thesis, London School of Hygiene & Tropical Medicine.

Tayeh A.T. and Cairncross S. (1991) The impact of water projects on the spread of dracunculiasis in part of Sudan and Ghana. In Wooldridge (ed.) *Techniques for Environmentally Sound Water Resources Development*. London: Pentech Press.

Tayeh A.T., Cairncross S. and Maude G.H. (1993) Water sources and other determinants of dracunculiasis in the Northern Region of Ghana. *Journal of Helminthology* (in press).

Tempalski B. (1991) The Spatial Distribution of Dracunculiasis in Northern Zou Province, Bénin. MA Thesis, Dept. of Geography, Hunter College, City University of New York.

Yellott H.L. (1990) Enquête sur l'Incidence de la Dracunculose dans Six Districts du Zou Nord de la République du Bénin. Cotonou: The Pragma Corporation.

Participatory methods in hygiene communication

Mrs Jemima A Dennis-Antwi

IN THE PAST educating for health in Ghana has been through the mass media, posters and didactic teaching. These methods do not take into consideration, the knowledge, values and skills already possessed by the learner. In providing the relevant information, target audience are often not:

- allowed to explore their own attitudes and feelings in relation to the subject.
- given the opportunity to utilise the information they already possess.
- allowed to explore any misconceptions and misinformation that they might possess on the issue.

As a result of this approach, target groups are not encouraged to develop the ability to make decisions about their own lives that will ultimately promote their health.

Participatory methodologies have been applied in disseminating information to the public on hygiene by the Health Education Project (HEP) of Kumasi Metropolitan Assembly (KMA). These methods ensure the involvement of the target audiences throughout the learning process.

The Project has a variety of participatory tools relating to water and sanitation including Three pile Sorting Cards, Flash Cards Series on diarrhoea diseases, worms, personal hygiene and Story- With- a- Gap . These methodologies have been extensively used in community and school education by health workers and teachers in the Metropolis.

The educational materials developed by the project has broad implications for health education in Ghana and elsewhere.

Participatory methods in hygiene communication

Access to sufficient safe water and adequate sanitation facilities are essential for public health, but having these facilities do not ensure an automatic improvement in public health.

Health benefits of clean water supply and adequate sanitation come only through proper functioning and use of facilities and often requires improved hygiene behaviour which can mostly be achieved through hygiene communication.

Defining the term *hygiene communication*

Several definitions of the above have been given by several authors. Boot,(1991) for instance defines it as " All activities aimed at encouraging behaviour which will help to prevent water and sanitation-related diseases (such as the various types of diarrhoea, worm infestations, skin and eye infections and vector-borne diseases)".

The focus of hygiene communication is on establishing links between facilities and practices with regard to: the use, care and maintenance of facilities; use of safe water in sufficient quantities; and the safe disposal of wastewater, human and solid wastes (Burgers et al,1988)

However the activities described above cannot be carried out unless one understands in clear terms the target audience's present behaviours, perceptions and priorities related to health problems. This is the first step in a participatory process between the facilitator and the target audience which goes a long way to ensure success of programmes.

Participatory approach to hygiene communication

The characteristic question that this approach focuses on is "How do I help people achieve what they want to achieve?" It therefore focuses on joint problem analysis and problem solving. This implies that the educator acts as a facilitator to create appropriate conditions to help solve the problems in a particular area. The objectives , content and methods are determined by the facilitator with the active involvement of the target audience through dialogues, discussions and meetings, among others. In providing the relevant information on a subject, the facilitator, takes into account the knowledge, values and skills already possessed by the target audience. By so doing the audience are:

- allowed to explore their own attitudes and feelings in relation to the subject.
- given the opportunity to utilise the information they already possess.
- Allowed to explore any misconception and misinformation that they might have on the subject without loss of face.

As a result, their self esteem is enhanced and this serves to empower them towards the adoption of appropriate hygiene behaviour.

The experience of the Kumasi Health Education Project (KHEP)

The Kumasi Health Education Project has since its inception in 1991 applied participatory methodologies in information dissemination to several target audiences including school children, mothers at antenatal and postnatal clinics and community members. These have been made feasible through a series of operational researches and training programmes for health education agents such as environmental health personnel, teachers and community health nurses in the use of the participatory methods in hygiene communication.

The project has also been able to produce its local participatory materials through a series of material development workshops in collaboration with the above health education agents and pretesting them using the target audiences. The finished products are then distributed to the agents to be used for hygiene education.

Some of the most popular methods employed by the project include:

Story with a gap

Its purpose is to demonstrate how a group can be engaged in identification of problems and planning water, sanitation and health activities. It is most appropriately used in the community. A typical example depicts two large posters one of which depicts a 'before' scene (a problem situation-water and sanitation) and an 'after' scene (a solution to the problem-sanitation at water source) as shown in Figure 1.

The target group is usually presented with the 'before' situation and comments are invited on what they see or to personalise the scene giving names. They are required to build the story up to a crisis point where something had to be done to improve conditions. Having established the 'before' situation as to how it happened, the 'after poster is shown and the group allowed time to discuss it. The question as to "What steps did the community take to change the condition from the 'before situation to the 'after' situation" is then asked and discussed. This generates a lot of ideas which can be systematically put together to help the group initiate a plan of action to improve their communities.

Three pile sorting cards

The purpose of this method is to help people develop analytical and problem-solving skills and the ability to reflect on causes and effects of sanitation related diseases. It also helps the facilitator to know the extent to which participants are fully aware of the positive and negative implications of a variety of situations shown to them.

The project has applied Three - pile Sorting Cards on malaria, diarrhoea and water for hygiene communication. A set comprises of about 15 cards each with a picture which could be interpreted as good, bad or in-between from the viewpoint of health, sanitation or water supply. The group(s) will be required to sort these cards into good, bad or in-between(those pictures which can be bad or good at the same time depending on one's judgement) and to justify their choice.

This method builds self confidence and has also helped in assessing the level of knowledge on some particular topics and to plan relevant programmes.

Flash card series

The purpose of this series is to increase knowledge levels of people in understanding systematically the process of disease causation. The flash card series developed by the Project for hygiene education include: Personal Hygiene, Diarrhoea Prevention, food hygiene and Prevention of Round worms. They have successfully been used in the schools, communities and clinics.

A series comprise of about 10 to 15 cards each with a picture relating to a particular disease. The facilitator stands in front of the group to lead a discussion on the topic by prompting or asking the target audience about what they see on a particular picture.

Based on their knowledge levels the facilitator provides the necessary information or facts to clarify any misconceptions or misinformation.

Evaluation of the participatory methods

An outcome evaluation research carried out by the Project on inservice training programmes for JSS/primary school teachers in the Kumasi Metropolitan Assembly on the applicablity of participatory health education materials clearly showed their effectiveness in leading to behaviour changes among the school children. The materials were mostly used during sessions on life skills. A similar research was carried out on the public sector health workers in the Metropolis who had been trained in the use of the materials. Again findings showed widespread use and effectiveness as reported by the health workers. The most popular methods were the flash card series.

The use of these methods are however not without problems. It usually requires time on the part of the target audience. For instance, mothers who attend antenatal or postnatal clinics are in a hurry to go back to their work and are therefore unwilling to wait for long periods. This means that the nurses have to repeat the process quite a number of times to different sets of the target audience. Secondly, it requires a suitable environment with adequate lighting to be used in the communities if it is in the evenings.

Based on these findings of the evaluation exercise, one can say that the use of participatory approaches to hygiene communication have positive implications for health education and can therefore be adopted by health education agents in other places or institutions to help people make informed choices about their lives as they see themselves as part of the process. Currently, Regional Health Education Officers in some regions in Ghana have expressed interest in adopting the participatory methods in their regions.[1]

References

1. Boot, Marieke T. (1991). Just Stir Gently: *The way to mix hygiene education with water supply and sanitation*.(Technical paper series;No 29). The Hague, Netherlands, IRC International Water and sanitation centre.

2. Burgers, L; Boot, M; Van Wijk-Sijbesma C.(1988).*Hygiene Education in Water supply and Sanitation programme:* Literature review with selected and annotated bibliography (Technical paper series; No 27). The Hague, Netherlands, Irc International Water And Sanitation Centre.

3. Srinivasan, L. (1990). *Tools for Community Participation - A manual for training trainers in participatory techniques*. PROWWESS/ UNDP Technical Series, New York.

[1] The above participatory tools were adapted from *Tools for Community Participation*, by Lyra Srinivasan.

Figure 1. Story with a gap on water and sanitation

The need for hygiene education

Susanne Niedrum

ONE OF THE main objectives of community water projects is to reduce water and excreta related diseases. A vital linking water projects to this objective is hygiene education (HE). But despite the many documents and policy papers on HE, there is little evidence that it is being taken seriously. Is this because decision makers are usually engineers? To compound this, more comprehensive efforts have often not been evaluated nor experience shared. The aim of this paper is to underline the need for integrated HE and to share the experience of the CARE Water Project in Rwanda.

The CARE Water Project is in the 7th year of its 10 year programme working with the rural population of 4 communes in north eastern Rwanda. Its objectives are to increase access to potable water and to help reduce water and excreta related diseases. It has set up, trained and supported 3 water associations with a total of 257 functioning water point committees. One of the associations is now completely autonomous and the others are following more or less closely behind. By the end of 1995 the Project will have supplied 70,000 people with potable water by spring protection and gravity systems. It has 4 integrated components: Community Management, which helps people to identify their need for water, to choose the technology, to position the water points, to construct, to manage and maintain them. Construction, which provides technical assistance and supplies non-local materials. Technical Training, which provides appropriate training to local water technicians and HE. The last two components were introduced within the last 2 years.

Lessons learnt

Like many water projects of the 1980's, the Project started life concentrating on its first objective of increasing access to potable water, hoping that the second, that of disease reduction, would follow of its own accord. In Rwanda although often polluted, water is abundant. The dispersed nature of the population and the fact that people prefer to live on the peaks of the hilly terrain often made it impossible for simple systems to bring potable water closer than existing sources. Of course none of the communities refused a new water system, but the fact that it often brought them no obvious advantage and that they had not taken part in key decisions resulted in people returning to their sometimes closer, polluted sources rather than paying the very low water fees. Neither did the new facilities lead to a reduction in disease as people continue to behave as they have always done. Water is contaminated before consumption, quantities have not increased and hygiene has not improved.

New approach

In the light of this experience, the Project has significantly modified its approach to concentrate on the objective of disease reduction. This approach is based on HE and on building up the capacities of the communities to analyse their own problems, to define their own priorities and take responsibility for planning, implementation and evaluation. For disease to be reduced, people must alter their behaviour and the water supply system must be sustainable. HE is paramount to both of these. Firstly, if people understand the relation between potable water, hygiene and disease, they are more likely to feel a need for potable water and its sustainability. If the community is thus the driving force behind construction and feels a sense of ownership and responsibility for the new system, problems with water fee payment should diminish. Secondly, HE can maximise the potential benefits of improved facilities by encouraging hygienic practices. The Project has high hopes that HE will also help to reduce Project extension costs spent on trying to resolve the water fee problem.

The HE staff will work in close collaboration with the community management staff with communities who already have water systems, with those who have systems under construction and with those who are yet unserved. In the latter case, water systems will not be constructed unless the community takes the lead.

This new approach necessitates a great flexibility with the Project. Construction has already slowed down significantly and in some cases may not take place as originally planned. Provision of water systems cannot be prerequisite to HE as it will not be possible for all those families who have participated in HE to be part of a construction scheme. In any case, HE alone, without improvement in water supply, should still be able to significantly reduce disease transmission. A great collaboration is required between the Project components, its partners and the community. To achieve this, regular meetings, workshops and training sessions are held at all levels.

Despite agreeing wholeheartedly with the philosophy, and despite the pressure on them from the other components and the Project manager, the engineers of the construction component still find it difficult to accept that they can no longer take the lead.

Staffing

The Project recruited a manager for the new HE component in January 1992. The post was open to men and women of technical, social and health backgrounds. The chosen candidate was a woman with a degree in community social work and 7 years experience with similar responsibilities. She works closely with her counterpart from the HF division of the Regional Health Authority. Shortly afterwards, 3 extensionists (2 men and a woman) were recruited. These people have completed A-level type studies in social work and have 3 to 5 years experience. The extensionists spent their first familiarising themselves with the community and undergoing training in the modes of transmission of water and excreta related diseases and how to avoid them. As the programme progressed they received training in educational skills, in training of trainers, in the design of educational materials and in the various participative information gathering techniques. Training continues as a regular part of the programme.

Data collection

Work started with a participative data collection phase to help identify needs and objectives. The first step was to hold an introductory workshop in each commune, the purpose of which was to explain the objectives of the component and to begin the participatory design process. Participants included the local authorities, representatives of the population and of the local institutions.

After this, data collection began in earnest. The following techniques were used: Visits to institutions including participative evaluation of any existing HE; population based survey of Knowledge, Attitudes and Practices (KAP) involving different members of the family and including observation; focus group discussions (FGD) with different social groups (men, women, adolescents, teachers, health workers); visits to all water points and discussions with users; household visits.

Programme design

Once the data had been collected and analysed, a 3 day preliminary design workshop was held between the Project and its partners to share the findings and begin design of the programme. To develop messages, participants identified risk behaviours corresponding to prevalent diseases. Attention was paid to why people behave in a particular way and the barriers to changing behaviour. Message groups were then developed, which advocate behaviour changes, which are realistic and build on existing values and practices. As all messages cannot be promoted at once, the participants chose priorities which counter the most harmful practices. For example, handwashing is considered to be one of the most important behaviours to encourage. The workshop then went on to outline relevant target groups, education methodology, canals used to pass the messages, the educational materials, training required and possible indicators.

Communication channels and methodologies

When asked with whom they would like to learn about hygiene, people said that they would prefer someone of the same sex but that they really did not mind, as long as the person had the necessary knowledge and understood their problems. Adult education is common and appreciated in Rwanda although under-resourced and organised in a rather adhoc fashion. The Project intends to use the existing channels of communication of churches, schools, adult education centres, health centres, water point committees and communal water technicians. They will be supported with training, supervision and logistics. Care was taken to look not only at people's formal responsibilities, but also the realities on the ground and their interests and motivation. Additional channels include community elected hygiene volunteers, the Project extensionists and other Project field staff, all of whom live in and often originate from the communities in which they work. The Project helped the communities to elect 2 volunteer extensionists per neighbourhood (a total of 480) and trained them in transmission routes of the disease, hygiene messages and in basic extension techniques. 88% of those elected participated in the training, which gives some idea of the enthusiasm for the topic. They will serve as a model to their neighbours, carry out limited extension work, and act as the "eyes" of the Project within the community. They are paid a per diem of US$ 1.4 for monthly training days.

Education methodologies adopted include public meetings, educational discussions, programmed learning, household visits, demonstrations, role play and discussions around testimonies given by people ill with a water or excreta related disease.

Throughout the planning phase the community often asked about educational materials. They were familiar with posters and flashcards and were also interested in receiving leaflets which they could refer back to and show their neighbours. It seemed that such material increased peoples confidence in, and the credibility of, the extensionist. The community will be heavily involved in the development of the material to ensure that it is culturally acceptable, comprehensible and interesting. It will include blackboard and chalk, leaflets, slides, posters and flash cards depicting various messages. Competitions will be held to develop stories and songs. All educators will receive a guide explaining all the messages. Testing will be done on whether it is necessary to use incentives such as soap, certificates or flags given to families trying new practices. Competitions will be held for the cleanest neighbourhood, the best water point and best constructed and clean latrine.

Pilot project

Unfortunately, just as the preliminary design stage had been completed (having taken one year), the war on the Uganda-Rwanda border flared up and progress was effectively stopped while the Project worked with the 200,000 displaced people who had arrived in the Project area. Now in June, the original work has resumed and the next steps are to go back to the population to present the results of the studies, to remodel the programme to their priorities, to develop programmes with the local institutions, to develop educational materials, to define targets and indicators and finally to produce a workplan. The Project will carry out continuous monitoring to provide immediate feedback on HE, on its integration with other components and the use of water supply systems. As it is always very difficult to accurately measure health improvements, the Project will focus on changes in behaviour for evaluation purposes.

A two year pilot programme will be run to enable various messages, education methodologies, channels and materials to be tested. This programme will be supervised in one commune by the Project and in the other three, by the HF division of the Regional Health Authority, with technical and logistic support from the Project.

Conclusion

The Project has learnt the hard way that if the Project is designed by an engineer and if the Project Manager is an engineer, construction is often prioritised. The omission of HE meant that people were neither informed about the dangers of non-potable water nor about the methods of reducing disease transmission routes. Water systems were built without full commitment from the community, which had serous repercussions for sustainability. In short, Project objectives were not reached.

The Project believes that HE and community empowerment are the inextricably linked steps towards making a sustainable success out of the Project, although they should not be looked upon as a miracle cure. This new approach requires the Project to be extremely flexible and highly integrated. Both HE and community empowerment are complex, sensitive and time consuming domains which require competent staff, sufficient time and sufficient resources. It is important to work within the existing structures and treat the community as a full partner in programme design and implementation.

Development of the HE programme has taken rather longer than expected, but some orientation and awareness-raising has been undertaken along the way and a sound participative base has been built for the implementation phase. It is encouraging that the communities have expressed such a strong desire to participate in HE.

Expenditure on HE during the first year design phase comes to just under US$ 55,000, having covered a population of 180,000. This represents just over 10% of total annual Project expenditure. Community management and technical training together account for another 15%, leaving the construction component with 75%. Expenditure includes salaries, benefits and per diems., training office equipment and supplies, communications, vehicle operation and maintenance (a jeep and 3 motorbikes). Note, should perhaps be taken that the 25% of the Project budget spent on "software" will probably be a lot more cost effective in terms of progress towards Project objectives that the 75% spent on hardware.

Copies of the Project Document and the various studies will be available at the conference. Comments are very welcome. The results of the pilot study will be published in detail in autumn 1995.

SECTION 4

INSTITUTIONAL DEVELOPMENT

Pricing water to recover costs

P J Barker

BANGALORE IS THE fifth largest city of India and is characterised by an average rate of population increase of approximately 4% p.a.

Selected census recorded populations:

	(m.)		
1901	0.228		
1941	0.510	Projected population	
1951	0.991	1995	5.2
1961	1.201	1999	6.1
1971	1.780	2000	6.4
1981	2.915	2005	7.8
1991	4.5	2010	9.5

Responsibility for the water supply and sanitation is vested in the Bangalore Water Supply and Sewerage Board. The water supply to the city is drawn from three sources:

1) Heseraghatta source - 18 kms. NW of the city. (22.5 mld.) 1884-1885
2) T.G. Halli source - 28 kms. W of the city. (143 mld.) 1933, 1951, 1964.
3) River Cauvary - 100 kms. SW of the city.
   Stage 1 - 1970-74 (135 mld.)
   Stage 2 - 1980-83 (135 mld.)

During Cauvary Stage 1 the entire city distribution system was remodelled and designed for a supply of 435 mld. inclusive of the 135 mld. Stage 1. (Based on a peak demand factor of 2.25 and consumption of 200 lpcd).

The history of the water sector in Bangalore is of one demand, (due to population growth, immigration and industrialisation) successively outstripping supply. The authorities have made a series of investments to fill this shortfall.

By 1967 the population had reached 1.5m. and per capita supply was only 100 lpcd. inclusive of industrial and commercial consumption. The response was sanctioning of Cauvary Stage 1.

Note that the application for World Bank loan assistance foundered due to the differences in opinion regarding 2500 litres monthly free allowance to consumers. In the event it was financed by the State Government, LIC and Debenture Loans.

Cauvary Stage 1 eased the supply position for two or three years from commission in January 1974 but rapid population growth by 1977 meant that supply was again inadequate with per capita consumption a maximum of 85 lpcd.

The response to severe shortages was essentially to duplicate Stage 1 in the form of Cauvary Stage 2 for a new increment of water, again of 135 mld. Improvements to distribution under Stage 1 accommodated this addition to supply.

Despite these additions to the system unremitting population growth has resulted in water delivery falling far short of the 140 lpcd. recommended by GOI guidelines.(*) A realistic estimate of availability put actual domestic per capita consumption at no more than 60 litres per day in early 1982 with an expectation that this would reach only 100 lpcd. after Stage 2 came on stream. The same analyst reported even this level of consumption would rapidly decline without further major investments (B.R. Nagendra, 1982). In order to avoid this deterioration the Board in 1985-86 started Cauvary Stage 3 at a cost of Rs.240m. and designed to bring an additional 270 mld. to Bangalore. The stage is due to be completed by mid 1993.

Metering and revenue

All house service connections and non-domestic connections are metered. Water supplied to domestic consumers up to 2500 litres is charged as per slab rate. The Bangalore City Council includes water charges in the House Tax and reimburses to the Board Rs. 7.50 per house for the free allowance.

Financial position

The income and expenditure of the Board over the period 1984-1992 was:

	Revenue (00,000's)	Expenditure	Surplus (+) Deficit (-)
84/85	1848	2089	-241
85/86	2127	2708	-581
86/87	2281	3060	-779
87/88	3582	3733	-151
88/89	3414	4018	-604
89/90	3530	4413	-883
90/91	3731	4410	-678
91/92	4049	4408	-359

The Board has incurred a loss every year since 1981. The revenue deficit (as a proportion of expenditure) was 20% for 89/90, 15% for 90/91 and 8% for 91/92. The Board stopped paying interest and principal to Government in March 1986. At present the Board has a deficit of Rs. 2.5m. even without repayment of Government loans.

In September 1990 the Chairman of the Board applied for permission to the State Government to increase rates which had last been revised in April 1987. These rates failed to provide sufficient revenue even to cover operation and maintenance costs and had resulted in large and growing budge deficits despite the cessation of repayment of Government debt. The main reason for failure to charge realistic tariffs seems to have been the desire of local politicians to curry favour with the electorate especially the poorer classes. In particular the Board's financial position was worsened by the escalation of establishment costs but also by the energy intensive nature of conveying water long distances from the points of extraction (up to 100 kms). Throughout the 80's the Board had been subject to a succession of energy price increases. The Board is heavily dependent on charges, over the period 84/85 to 89/90 between 85% and 93% of total revenue came from water rates and water meter service charges. Other receipts was a minority item ranging from 7% to 15%.

This brief review establishes the serious and escalating adverse financial situation faced by the Board and the importance of water charges as a revenue generator. It is essential that water charges should cover the full cost of production and delivery if the Board is to meet its mandatory 'no profit - no loss' obligation. In particular given the importance of energy and maintenance costs water rates must be revised so as to compensate for increases in these two items.

Revision of the tariff structure

The objectives of the tariff structure are:

i) To recover the cost of water produced and delivered. This enables the organisation to meet its financial obligations, to provide a sustainable service and to signal to consumers the need to avoid waste and to provide an incentive to conserve water. Finally the tariff should allow the Board to make provision for future necessary increments to supply.
ii) To ensure that payments are consistent with local ability to pay bearing in mind the various income levels and the need to meet public health objectives. In this latter regard the cost of basic needs water is particularly pertinent.

The first objective may be referred to as an economic efficiency aim and the second as a distributional (or social equity) aim.

Because water supply in the immediate future is a mixture of different vintages produced from past and current investment each producing at different costs the tariff should reflect these various costs. The system recommended is to:

a) Calculate the cost of 'new' water (i.e. Cauvary Stage 3) to find its Average Incremental Cost.
b) Determine the AICs for each of the past investments i.e. for each vintage of water.
c) Calculate the weighted average cost of water currently consumed according to the contribution made by each vintage in total production.
d) Distribute the weighted AIC across the income classes in such a fashion so as to recover costs and to simultaneously fulfil the equity requirement.

The AIC is defined as Present Value of Total Costs divided by the Present Value of Output (where output is in effect acting as a proxy for benefits). AIC was calculated for each vintage or tranche of water. These were:

Rs. .45 for pre-Cauvary water   (165 mld)
Rs. 1.70 for Cauvary Stage 1 water   (135 mld)
Rs. 2.70 for Cauvary Stage 2   "   (135 mld)
Rs. 5.82 for Cauvary Stage 3   "   (270 mld)

Total   705 mld

These costs were weighted according to their individual contributions in the total supply of 705 mld after completion of Stage 3. On this basis the weighted AIC of water from all existing sources is Rs. 3.17 (say Rs. 3.2) per 1000 litres.

It was estimated that making allowance for failure to use the full potential of water and water losses would reduce the total actually available for consumption from 705 mld to 541 mld. Thus the average weighted price sufficient to recover costs becomes:

Rs. 3.17 x 705 / 541 =   Rs. 4.13 per cu. meter
                   Say   Rs. 4.2 per cu. meter

The proposed tariff

A simple tariff was designed to achieve the aims of efficiency and equity. Simplicity, fairness and cost recovery are essential elements of the tariff (Table 1).

Annually this is Rs.1017.m. Currently, the Board levies a sanitation charge equal to 25% of the water charge. It would actually like to charge 40%. Using the former figure generates (Rs.1017 x 1.25) = Rs. 1271 m. per annum.

The annual total cost of water production calculated on the basis of weighted AIC = Rs. 4.2 x 705 Mld x 1000 x 365 = Rs. 1080.7m. Thus the proposed tariff would generate sufficient funds to cover the full annual cost of water production. An increase in sanitation charges would allow a more generous contribution to future systems expansions.

The affordability of the proposed tariff

On the basis of data relating to consumption and discussions with Board officials with accumulated knowledge of local socio-economic conditions the above data was assembled (Table 2).

The most vulnerable group are the slum dwellers who have an estimated family income of Rs. 1700 per month. Under the old tariff arrangements this group had re-

Table 1

Monthly flow (in litres)	No. of Connect's	Ave. flow (1.p.m.)	Total (MLD)	Tariff/1000 1.	Rev/Rs./m. (mills
DOMESTIC					
0 - 10000	60000	10000	20	Rs.1	.6
10 - 25000	90000	20000	61	2	2.7
25 - 50000	60000	40000	97	3	5.1
50 - 100000	18000	90000	54	4	4.95
100000 +	2000	165000	110	3	24.3
Pub. Taps.	5300		102	1250 tap/m.	6.6
NON DOM.					
0 - 10000	4000	7500	1	4	.12
10 - 25000	2200	13600	1	5	.15
25 - 50000	2000	45000	3	6	.54
50 - 100000	1300	92000	4	8	.095
100000 +	1200	2200000	88	15	39.6

GRAND TOTAL = 84.75 per month

Table 2

Econ Group	% of Pop.	No. of Families	Ann. Inc. Income Rs./Fam.	Monthly Income Rs./Fam.	Ave. Consumption /month./family cu.m (a)	(b)	Total Cons. Mnth. 000s
Slum	7	60000	20000	1700	3.75	9.75	585
V. Low	33	300000	35000	3000	5.25	12.00	3600
L. Mid.	30	270000	50000	4200	10.5	15.00	4050
Up. Mid.	20	180000	75000	6300	13.5	18.00	3240
High.	10	90000	150000	12600	17.25	19.5	1755

(a) Monthly Family Consumption Pre-Cauvary.

(b) Monthly Family Consumption Post-Cauvary. The sum of the 5 classes of consumption = 13,230,000 cubic meters per month which is equivalent to 441 MLD and is thus within the capacity of the Post-Cauvary system.

The average weighted cost of Post-Cauvary 3 water was as reported found to be Rs. 4.20 per cubic meter.

ceived 'free water' an arrangement which had contributed to the Board's poor financial position. The estimated monthly consumption (post-Cauvary 3) is 9.75 cubic metres. Thus at the proposed slab the poorest would pay Rs. 1 per cubic metre or Rs. 9.75 per month. Even allowing for a 40% sanitation charge this would amount to less than 1% of average share family income. Such a burden for a superior service is well within accepted guidelines for an economically efficient and socially equitable system.

It is interesting to note that even if the poorest paid the full AIC and sanitation was set at 25% of the water charge the poorest would pay only 3% of family income per month.

(*) Recommended at the Conference of Secretaries, Chief Engineers and Heads of Implementing Agencies. Mysore 1989.

Project Formulation and Appraisal Techniques Applied to a Water Supply Case Study. Unpublished Postgraduate Thesis, Anna University, 1982.

Private sector involvement

Erich Baumann

TO ESTABLISH THE construction, drilling, manufacture and maintenance capability in the country it is necessary to follow a clearly defined policy framework under which a transition from the present *"construction by GWSC projects"* to a future *"construction by private sector"* is realized. Standardization and clearly defined regulations are the essential precondition to make this step. The private sector can only successfully operate if GWSC, ESAs and NGOs coordinate their activities and policies in such a way that all private enterprises have a chance to obtain work in a free competitive environment. Procurement practices have to change from International Competitive Bidding to Local Competitive Bidding. A key factor for the development of a strong private sector is the coordination of sector activities at regional and district level.

Technical specifications

Standardization and clearly defined regulations are essential preconditions to change the sector. Provision of goods and services through the private sector has to be achieved through a clearly defined technical regulatory system. GWSC will have to standardize the technologies used. Standard designs would be established for (1) Latrines, (2) Hand dug wells with cement block lining and with insitu lining, well head design for handpump, (3) Boreholes, 4", 5" and 6" in diameter, well pad design, (4) Handpumps, direct action, deep well, (5) Small piped systems, solar-, solar/hybrid-, diesel/electric, electric pumps, storage tanks, piping, standpipes. For each technology complete technical specification, including drawings, bills of quantities, production procedures and quality control requirements, will have to be worked out. Similarly, rules and guidelines for operation and maintenance will be formulated. These technical specifications would be the basis for any local or international competitive bidding.

The decision to standardize on equipment has been endorsed by the GOG (PNDC Committee of Secretaries). GWSC will have to build the capacity in a Technical Department with the necessary personnel to maintain and administer the standards and specifications.

The private sector participation in the formulation of standards would be achieved by inviting associations of drilling companies, handpump suppliers and/or construction companies to take part in the elaboration and the review of the standards. These associations would have seats allocated in the standards review committees.

Pre-qualification of suppliers

The next step will be the identification of suppliers and/or manufacturers that are capable of producing the products/services as specified. Information will be gathered on: a) Technical ability; b) Infrastructure/equipment availability and type; c) Financial capability; d) Past performance; e) Internal quality control. If the above criteria are satisfied a works inspection will be carried out by a inspection agency to verify the data received. A trial order will be placed with the prospective supplier. During the execution of the trial order visits to site/factory will be made to ascertain whether the necessary technical, managerial skills are available and the quality control system is in place. These visits will reveal where the contractor is deficient and what specific training may be required. A supplier/contractor that has passed the above procedure may be registered as a *Certified Supplier*. Only suppliers that have reached this status will be eligible for participation in bidding on subsidized contracts. The performance of the contractor will be reviewed on an annual basis. Defaulters will lose their license to operate. Depending on the nature of the work, pre-qualification may be done on national level (drilling contractors, handpump suppliers, etc.), on regional level (hand dug well and latrine contractors, maintenance contractors, community animation contractors).

Manufacture and distribution of handpumps

It can be estimated that the market in Ghana will eventually be 2,500 to 3,000 pumps per year. The local manufacturers could cover approx. 50 to 60% of this market, i.e. about 1,200 to 1,800 pumps per annum.

The investments required for equipment and tooling as well as the high working capital demand constrain the choice of companies. Small scale and cottage industries would financially not be able to undertake such a venture and would not have the managerial ability for sustained successful production. Economic conditions in Ghana have considerably improved over the last 4-5 years. Under the Ghana Investment Code it is possible to obtain quite favourable relaxations on import restrictions and tax and duty exemptions. However it is still an uphill battle for local industries trying to survive and to compete against imported products. Handpumps can be imported duty free, and (for instance) manufacturers in India profit from export subsidies.

Realistically, local manufacture is not always feasible. Ghana, even though theoretically large enough to sustain about 2 - 3 local manufacturers, will probably in the near future have to satisfy herself with one manufacturer for direct action pumps. In the meantime it is essential to strive towards a more comprehensive understanding of supply of handpumps for CWS/S Projects. If the pumps are supplied through the local, private sector as turn-key installations (including: supply, distribution, delivery to site, installation, training of village mechanic, provision of after-sales services) the locally added value is quite considerable.

Recommended actions to improve manufacturing/distribution capacity

Supplier (importers) and even more so local manufacturers however have to import/produce pumps in reasonable quantities. Raw material cost is about 50 -60 % of the sales price requiring high working capital. The establishment of after-sales services as part of the process of pre-qualification requires additional substantial capital inputs. GWSC would order an anticipated two years demand of standardized pumps. It would call for LCB for collective orders with pre-qualified suppliers. GWSC would pay an advance of 70%. The produced/imported pumps would be inspected in the premises of the manufacturer/supplier by an independent inspection company and sealed. During the process of project execution the regions/projects would release the delivery of small quantities for installation. After completion of installation the supplier will be paid 20 % of the price by the community. GWSC will retain 10% over the guarantee period of 12 months.

Distribution and installation of handpumps

Contracts for the supply of handpumps will include the provision of hardware as well as related services. The suppliers will be required to establish a comprehensive network of Regional Dealers and Spare Part outlets. These regional dealers and their area mechanics will be the backbone of the repair services. The table below indicates the division of functions for distribution and installation of handpumps:

GWSC
- Selects 2 - 3 Pump Types for Standardization
- Defines Specification
- Pre-qualifies Suppliers

CWS/S project (Bilateral and NGOs)
- Plans the project
- Purchases Pumps
- Organizes Quality Control
- Service Contract with Supplier
- Supervises the works

Supplier/manufacture
- Imports/manufactures the pumps
- Custom clearance
- Stocks Pumps and Spares
- Appoints Regional Dealers
- Distributes Pumps to Regional Dealers
- Training of Mechanics, Area Mechanics, Installation Crews
- Marketing of the Pumps
- Liaison with Manufacturer
- After-Sales Services

Regional dealer
- Distributes Pumps to Site
- Pump Installation
- Training of Village Mechanics
- Stocks and Sells Spare Parts
- Provides Repair Services
- Liaison with Area Mechanics

Area mechanic
- Makes the annual Inspection
- Repairs Pumps and sells Spare Parts on commission
- Assists Regional Dealer in Installations
- Liaison with Community

Watsan pump mechanic
- Management of O&M
- Minor Repairs
- Collection of Inspection fee

The criteria for pre-qualification compels the suppliers to gear up to these requirements. The cost of the services provided by the private sector will affect in the cost of the handpumps. The installed price of a handpump would be approx. 1.6 to 2 times the FOB price.

Maintenance and Repair Services Community management of O&M will not relieve the government of its continued involvement in rural water supply. It might however be possible for the government to assign some of the duties to the communities and the private sector.

GWSC will have to formulate a general maintenance strategy, for newly developed facilities as well as for existing facilities. The strategy needs be drafted in such a way that it would cater in a practical manner for the transition of the present O&M system to the new CWS system, preserving the already existing infrastructures, keeping the present O&M structures operational and thereby making best use of the personnel resources in GWSC. This strategy would be based on the following:

- Acceptance that O&M needs to be subsidized also under community management. The strategy will include the establishment of a Rehabilitation Fund that covers certain aspects of O&M (well maintenance, major repairs, etc.). Initially donors would inject seed money to get the fund started. The GOG would gradually take over the financial responsibility to replenish the Rehabilitation Fund by using either its own funds or counterpart funds generated with sector funding by ESAs.
- Assessment of the existing resources. GWSC Maintenance Units are not dissolved but gradually changed and incorporated into the new CWS system. Eventually privatization of the services provided by these units will be encouraged.
- Standardization of Handpumps. Involvement of the private sector for O&M and the provision of spare parts make it mandatory to keep the number of pumps to a minimum.
- Establishment of a continued annual preventive maintenance and inspection service. Communities, when they assent to obtain a water facility, sign an agreement that they will have the facilities inspected by a

GWSC approved private mechanic on a regular basis. During the inspection the fast wearing parts are exchanged. The communities will have to pay a fixed rate for this service to the mechanic. This will safeguard that the investments made by GOG are used and maintained properly.
- O&M Support by the District Assembly would be permanently provided by the O&M Technician of the District Water and Sanitation Team (DWST). He is responsible for advising and motivating the communities. He is the link to private mechanics. He will be responsible for maintaining the district data base of all water facilities. The O&M Technician would be part of the team that scrutinizes the applications to the rehabilitation fund.
- Any other repairs and additional spare parts which become necessary between the inspections will have to be paid by the communities.
- Back-up for major repairs that are well outside the technical and financial capability of the communities would need to be ensured by GWSC. Borehole redevelopment would have to be organized. Presently GWSC Maintenance Unit (MU) has the capacity to maintain all the boreholes in the country. However it will be necessary to streamline and eventually privatize the MU operations.

In case of a major repair or borehole redevelopment the community will have the possibility to apply for financial assistance to the district assembly. Precondition to be eligible for any assistance from the rehabilitation fund would be that the community had all the annual inspections carried out. The cost sharing arrangement will include that the communities will have to pay a reasonable percentage of the cost.

- Introduction of the CWS system in some selected districts only. GWSC will carry on with the present centralized handpump maintenance system under the Maintenance Unit in the other regions for some time before changing also over to the CWS system with regular inspections. Tariff payment would stop and the MU would charge the communities the recommended price for the inspections. The communities in these regions would be told that a change in the maintenance system is imminent and they would be educated towards this goal. Since they have been used to paying regular tariffs it should be quite easy to motivate them to cooperate with the new maintenance structure. This approach would allow GWSC to implement a nationwide maintenance strategy in steps without compromising on the present pumps.
- Identification of what services can be privatized. Franchising of some of the services to NGOs or private companies which are acting on behalf of GWSC will be considered.
- Assistance to the private sector to develop a network that can provide the spare parts and services. Pre-qualification of suppliers demands that after-sales services are established in the regions and districts during the supply of the initial pumps/equipment. This will guarantee that communities have spare parts and trained mechanics in the vicinity.
- Training of community pump mechanics, area mechanics and O&M Technicians will be done by the suppliers under supervision of GWSC. District and GWSC personnel will need to be trained on all technologies.

Private sector involvement in O&M

Suppliers of handpumps would be compelled to set up their own after-sales service network. Thus the private sector would be responsible for the supply and distribution of spare parts. Spare part distribution on its own is not economically viable. It has to be tied in with the supply of new equipment. If the renewal of pre-qualification depends on the continued availability of spares in the districts at reasonable cost, suppliers will ensure the supply of spare parts. Regional sales centres equipped with the necessary tools and spares will be set up.

Area mechanics are small enterprises with part time activity to inspect, maintain and repair handpumps. Even though the regular inspections would provide a basic work load, the economic base of such a job is not sufficient to support a mechanic full time. The suppliers would be advised to appoint small entrepreneurs who are already operating in a related field (car/motorcycle mechanics, household articles repairers) as their district representatives.

In order to perform the (preventive maintenance) inspections the area mechanic would need to be certified by GWSC. For this he would have to undergo the specified training. The training of area mechanics would be left to the supplier under the guidelines for pre-qualification.

Spare parts distribution

The pre-qualified suppliers will have to set up a network of spare parts outlets. The distribution network could have the following appearance:

- The national supplier keeps fully comprehensive stocks of spares in his central store.
- In each region the regional dealer will keep adequate stocks of spare parts. The financial risk will be with the national supplier. The regional dealer will pay a deposit of (let´s say) 20% of the value of the spares that he has in stock. The margin for the regional dealer could be about 30% of the sales price. The quantity of spare parts in stock would have to be sufficient to cover at least 80% of all breakdowns. In the case that the components would not be in stock with the regional dealer, he should be able to order the parts within one week from the national supplier. The regional dealer would sell the spare parts either directly to communities or through the appointed area mechanic. The bulk of the spare parts sales would be parts that are to be replaced during the (preventive maintenance) inspection. This would allow the regional dealer to plan fairly closely the annual turnover on spares. These planning figures

would help the national supplier to establish the demand for spare parts on national level.
- The area mechanic is the principal outlet for spare part sales. He makes the annual in-spections, replaces the fast wearing parts. He would have a margin of (let's say) 20% on all spares. He would stock only the parts he has to replace during the inspection. He would have to order from the regional dealer any other spare part required for repairs outside the scheduled maintenance. The distribution network would need to be set up in such a way that (95% of the) spare parts could be available at the area mechanic in less than a week.

GWSC would prepare a list of recommended spare parts sales prices for all the standardized pumps. This list would be reviewed and agreed on annually together with the pre-qualified suppliers/manufacturers. The price list would be published as recommended prices so that the communities would know how much the spare parts cost.

Legal module for environmental protection

Dr Mrs V Hemalatha Devi

Quality of life encompasses many parameters. Environmental quality is the most important of these parameters. There is a great deal of awareness in India for the urgent need to preserve the ecological balance in the process of development and industrialization. Unplanned activities of man are degrading the environmental quality, threatening the web of life. Technological advances can help in improving the degraded environmental quality.

Past experience shows that technology alone cannot handle the situation. Governmental, institutional and social actions are desired to maintain the quality of environment. Legal sanction and action are required for the purpose.

The present paper covers the problems being faced by developing countries in maintaining environmental quality with the assistance of legal sanctions. A legal module for environmental quality is presented for effective implementation of the provisions of environmental legislation.

Environmental legislations

Regulatory legislation to control environmental pollution is necessary since the rule of law must protect the rule of life. The Indian Constitution provides under Articles 47, 48, 48A, 49 and 51-A(g) for the protection of environment. Salient legislations in this direction enacted by the Indian parliament are The Water (Prevention and Control of Pollution) Act, 1974; The Air (Prevention and Control of Pollution) Act, 1981 and The Environment (Protection) Act, 1986.

Aspects of implementation

Environmental issues have no administrative, socio-cultural or political boundaries. Inspite of the statutes being available, there are several problems in effective implementation. These problems are being faced by almost all developing nations.

Success in achieving environmentally sound development will depend on the extent of co-operation that can be achieved between government, its subsidiary agencies, voluntary groups, financial institutions, corporate groups in the public and private sectors, education and research bodies, professional societies, religious and cutural institutions.

- The industry in general is not prepared for being asked to control or stop pollution of the environment. It assumes that its role is limited to producing goods and treating the effluents or emissions is not its duty. The industry also harbours the notion that investment in pollution control is unproductive.
- Industry's money power is able to secure the services of better lawyers than those who typically represent environmental agencies and government departments.
- Environmental considerations are not being given their place in decision making as regards planning and location of new industries and expansion of existing ones. Economic and other considerations outweigh environmental aspects.
- It is a fact that closure of industries may bring unemployment, loss of revenue, but life, health, and ecology have greater importance to the people.
- There are two schools of thinking in developing effluent/emission standards in developing countries. Some people believe that under nourished people require pollution limits lower than those of other countries with far better nourished population. Others believe that developing countries should have lenient standards to encourage industrialization. This is creating problems in prescribing standards in most of the countries.
- Lenient pollution control standards made applicable to old industries located near urban areas came in the way of effective implementation of control laws.
- In big countries, there exist different effluent standards prepared by different agencies of the government, resulting in problems in implementation. In the absence of unified set of standards, legal problems are being created.
- The prosecuting pollution control boards often feel frustrated in being unable to achieve their goals on account of technical legal problems.
- The concept of 'conditional consent', being adopted by pollution control boards is tantamount to giving a license to pollute. Once the consent is given, the board find it very difficult to monitor or take action if conditions laid down in the 'conditional consent' are not complied with by the industry.
- Frustrated at the lengthy court procedures, the boards find it convenient to drop the cases or as an alternative, formulate a more lenient policy towards the industries.
- The pollution control boards can do nothing once the case is sub-judice and the environment degradation goes on unchecked in the mean time.
- The laws of evidence are tortuous with consideration to the infringement of environmental laws.
- The litigation process is very slow; many years pass between prosecution and decision allowing for unabated pollution in the interim period.
- Lack of data regarding the health effects and other impacts on quality of life does not convince the judges to award the punishment prescribed.

Legal module

The law-men should engage themselves in finding out how far the laws are sufficient and to what extent these laws require reinforcement, re-orientation and modification. There is no dearth of environmental protection laws, but a firm hand is needed to implement them. Several actions are required to sharpen the teeth of the statues of pollution control.

- A comprehensive "Environmental law code" should be enacted envisaging specific liability - civil and/or criminal of individual or individuals.
- Instead of dragging questions of environmental violation to a court of law through prosecution, it is advisable to provide the enforcement authority themselves with powers of imposing prohibitive penalties so as to make the cost of violation more than the cost of observance.
- In the interest of speedy and effective remedial justice, separate environment courts should be established with simplified procedure. The responsibility to disprove pollution should lie heavily on the accused.
- Public interest litigation for protection of the environment should be permitted in view of the wider social interest affected by environmental pollution. Public interest litigation should be institutionalized by providing statutory recognition.
- The polluter pays principle must form part of the legal machinery.
- Through legal means, rehabilitation of the victims who have become disabled due to pollution, must be provided.

To solve the problem of ineffective legal management of environmental quality, a permanent solution is to be found. A legal module can be developed by every country to handle the situation.

The module shall consist of separate units at national level, state level and district level. These units shall handle environmental litigation at national, state and district level respectively. The district level unit can solve most of the cases. These units shall be headed by the immediate retired judges of the supreme court, high court and district court respectively. The legal module will be different from pollution control boards and will be an independent government funded system. Advisory committees consisting of members from industry, engineers, technologists, medical specialists, pollution control boards, general public and women are to be formed to provide advice to the units at each level. The legal module is to be incorporated through a legislation so that the implementation aspect of legal statutes can be taken care of.

The proposed module can be extended to the international level by involving U.N.O. Although the U.N has no power of its own to be effective, it can make all the nations come to an agreement to create a legal module which can take effective steps to maintain global environment.

Thinking things through

Duncan Morris

THIS PAPER IS about making the provision of basic infrastructure more responsive to the needs of poor people in both urban and rural areas, more affordable to both the users and the providers, and more sustainable in the long term.

It suggests that these ends can be achieved by accepting three complementary basic principles:

- a limited role for government and the public sector;
- staged development to affordable standards at each stage; and
- optimum use of locally available resources;

and thinking through their implications concerning the formulation of policies and programmes for infrastructural development.

A limited role for government and the public sector

In most developing countries the provision of basic infrastructure is in the hands of (central and local) governments with international agencies and NGOs in support. There have been many attempts at community participation but all too often these have been seen as a way of keeping down costs and shedding maintenance responsibilities rather than making programmes more responsive to people's needs.

It is probably fair to say that the idea of asking communities to participate in what are essentially government projects has not worked very well. There is a strong case for a new approach to the provision (and maintenance) of basic infrastructure in which governments:

- restrict their direct interventions to the provision and maintenance of publicly owned and used infrastructure and utilities, through their normal programmes of *public works;*
- encourage and help people to identify and meet their basic needs, with a minimum of government involvement, through programmes of *community works;* and
- *draw a clear distinction* between public works and community works.

This distinction should be based not so much on who pays for the works, as on who initiated them and who owns, manages, uses and maintains the assets created.

Public works are usually initiated and financed by government agencies and executed according to pre-determined programmes. The assets created are owned, managed and maintained by the responsible agencies and used by the general public which pays for them either directly (through user-charges) or indirectly (through taxation).

There are three main ways of limiting the role of government and the public sector in public works:

- by decentralizing and delegating decision-making and local resource mobilization to the appropriate level of government (according to the significance of the infrastructure) but allocating central government support on the basis of national guidelines and criteria (thereby limiting the influence of local politics);
- by making as much use as possible of the private sector (consultants and contractors) to prepare and implement public works projects, and also to manage, maintain and where necessary operate the assets created; and
- by simplifying and streamlining the regulations and procedures relating to the provision and maintenance of basic public infrastructure, such as rural feeder roads.

Public works are by definition not community-oriented. Thus, community participation in public works should be essentially consultative (through participatory planning) rather than active (during implementation). Indeed, people who work on public works projects do not necessarily benefit directly form the assets created. Thus, they are workers and should be remunerated in full for their labor, a reasonable proportion being in cash.

Community works are quite different. They are initiated by clearly identifiable communities or common-interest groups (or even individual households) for the mutual benefit of their members. They are usually prepared and implemented, at a pace determined by the beneficiaries, with the help of *facilitating/implementing agencies* which provide technical, material and financial assistance. The assets created are owned or at least managed, used and maintained by the beneficiaries, who have to decide how to share the costs (and in some cases the benefits) amongst themselves.

So, rather than talking about community participation in government projects, we should be talking about government participation in community projects and this should be essentially:

- *promotional,* eg. publicizing the benefits of sanitation and telling people how they can get help to build a pit latrine;
- *enabling,* ie. encouraging people to help themselves and providing indirect support ; and

- *regulatory*, through the granting of planning permission and technical approval (but this should be made as straightforward as possible).

This does not relieve governments of their responsibility to support community works (which should be subsidized to at least the same extent as public works) but it does mean that community works should be de-politicized and, ideally, that all resources (from government, donors and the beneficiaries themselves) should be channelled through *independent executing agencies* (like the AGETIP-type agencies in Francophone West Africa). This offers several important advantages, notably:

- *responsiveness*, in that people should be able to take their problems (as they see them) directly to the independent agencies without going through the various tiers of local government;
- *flexibility*, in that (i) the agencies should be able to accept beneficiaries contributions in cash (with access to credit) or in kind (but costed), and (ii) the levels of subsidy and priority ratings allocated to different types of projects can be varied (according to pre-determined rules) to reflect government and donor policies and the needs of the beneficiaries (as perceived by the agencies);
- *efficiency*, in that the executing agencies being independent can short-circuit cumbersome government regulations and procedures, and employ NGOs or the private sector as facilitating/implementing agencies.

In short, it is recommended that:

- community works be supported to at least the same extent as public works but with the support being indirect and pooled (albeit with the possibility of favoring specific sectors and activities through higher priority ratings and subsidies); and
- the beneficiaries become the clients, with donors and governments placing more reliance on their innate commonsense than they have in the past.

Staged development to affordable standards at each stage

One of the major drawbacks of conventional infrastructural development is the practice of building infrastructure to an unnecessarily high standard in the first place and them upgrading it to its ultimate standard in one or two large steps. This practice is reinforced by master planning exercises which, while they are useful for setting ultimate targets, are seldom realistic as regards the resources available and time needed for implementation.

The simple fact is that this is not an appropriate approach to provision of in low-cost, basic infrastructure where coverage needs to be maximized and risks and costs minimized. This requires that infrastructure provides the minimum acceptable level of service initially and is upgraded in several small steps to affordable standards (in line with demand) at each stage.

The main implications of staged infrastructural development may be summarized as follows:

- infrastructural development policies and master plans should make provision for staged implementation or interim developments to appropriate, affordable standards;
- effective participatory planning procedures are needed at all levels of government;
- staged development provides a mechanism whereby community assets can be upgraded and transformed into public assets;
- the size and timing of each stage/step should be based on:
  - changes in significance, function and ownership;
  - technical thresholds;
  - demand (or use); and
  - the availability of resources;
- when the significance, function and ownership of infrastructure change, upgrading should be the responsibility of the new owner (the opposite of conventional practice); and
- as mentioned before, decision making and local resource mobilization should be decentralized and delegated to the level of significance of the infrastructure, but central government resources should be allocated according to national guidelines and criteria.

Optimum use of locally available resources

Conventional infrastructural development tends to be capital-intensive and heavily dependent on imported technologies, skills, materials and equipment. This makes it very expensive (especially in foreign exchange) but with relatively little money being spent in the locality of the work.

Such an approach may be appropriate for large-scale, technically complex works, but it is certainly not appropriate for the provision (and maintenance) of low-cost, basic infrastructure, where much more use should be made of locally available resources, especially labour. However, in many African countries there is still considerable resistance to local resource-based, labour-intensive technologies which tend to be seen as backward, slow and second-rate. This is not so; the aim is to produce infrastructure of comparable quality but at a lower cost than that produced by conventional capital-intensive methods.

Optimum use of local resources implies cost: effective substitution of labour for capital and local resources for imports, and it is based on rational technology choice, as illustrated in Figure 1.

Adopting a local resource-based labour-intensive approach to basic infrastructural development offers several important advantages, notably:

- making external assistance go further (provided that governments regard external assistance as supplementing rather than replacing their own resources);
- reducing dependence on scarce foreign exchange and expensive (but often unreliable) imports;

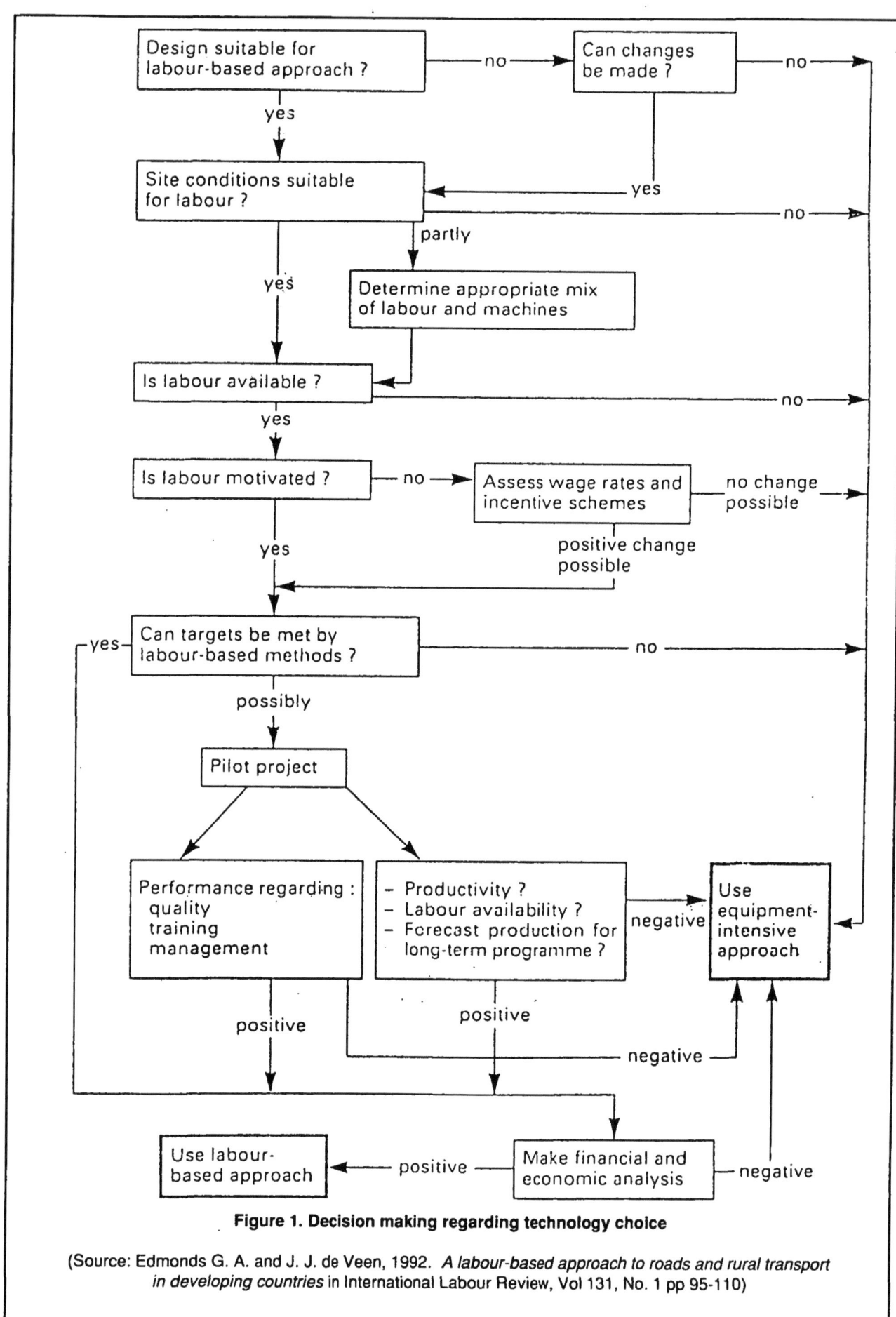

Figure 1. Decision making regarding technology choice

(Source: Edmonds G. A. and J. J. de Veen, 1992. *A labour-based approach to roads and rural transport in developing countries* in International Labour Review, Vol 131, No. 1 pp 95-110)

- generating gainful employment from public works and making community works more affordable; and
- in the longer term, bringing local resource-based, labor-intensive methods into the mainstream of infrastructural development which should result in substantial and sustainable increases in employment.

However, the realization of these advantages is dependent on certain conditions being fulfilled:

- there needs to be a framework of explicitly supportive policies and reinforcing measure ;
- the concept of rational technology choice and the application of local resource-based, labour-intensive technology need to be introduced into the curricula of universities and technical colleges;
- advisory support, information services and training should be available to all interested parties government and public sector, private sector and NGOs and workers' organizations and cooperatives;
- actions taken in the short term should lay a foundation for gradual expansion (in line with capacities to identify, prepare and implement projects and to maintain the assets created), and under no circumstances should they be allowed to jeopardize the prospects for expansion in the long term.

Conclusions

Each of the three principles helps to make the provision (and maintenance) of basic infrastructure more responsive, affordable and sustainable. However, it is only when all three are applied in a concerted way that their full impact will be felt.

The most important implications concerning the formulation of policies and programmes for infrastructural development may be summarized as follows:

a. A clear distinction needs to be drawn between public works and community works, and this should be based on who initiated the works and who owns, manages, uses and maintains the assets created.
b. Decision making and local resource mobilization should be decentralized and delegated to the appropriate level of government (according to the significance of the infrastructure) but central government resources should be allocated according to national guidelines and criteria (to limit the influence of local politics).
c. Participatory planning is needed at all levels of government.
d. The bureaucratic regulations and procedures governing the provision and maintenance of basic infrastructure need to be simplified and streamlined.
e. Community works should be supported to at least the same extent as public works but indirectly and without the support being tied to specific sectors or activities so that the beneficiaries are able to seek help with what they see as their most pressing problems.
f. Much more use should be made of the private sector (especially as regards public works) and NGOs (especially as regards community works).
h. Local resource-based provision and maintenance of basic infrastructure can create substantial and sustainable increases in employment and this should be taken into account when planning infrastructural development.

Acknowledgement

This paper is based on eight years experience as the International Labour Organisation's Regional Adviser for Labour-intensive Works in Africa and the author wishes to thank the ILO for allowing him to write and present this paper. However, the views expressed in this paper are the responsibility of the author alone.

Improving water supply through privatization

Professor S Mustafa

IN MOST DEVELOPING COUNTRIES, Government policies on water supply, if any, fail to take cognisance of the role of Private Water Vendors (PWVs). Often, political or social considerations constrain most public water corporations from charging appropriate water rates that will enable them to recover even operating costs. Thus, despite huge government investments over the years in the water sector, Public Water Corporations in Nigeria due mainly to low water tariffs that they charge, experience difficulty in maintaining and operating their systems. This situation has resulted in continued deterioration of treatment plants and distribution networks, leading to inadequate and poor service as enough funds could not be generated to replace broken down parts or expand existing systems. The current world wide economic recession has caused most developing countries particularly to experience, in differing degrees, deteriorating per capita income growth, stagnating if not declining government revenues and serious balance of payments and debt servicing difficulties. These have led to drastic cut backs in investments and slow down in ongoing projects. To deal with the problems of broken down plants and expansion of existing services, both Federal and State Governments in Nigeria resorted to external loans to finance water projects.

Water tariff is too low in all State Water Agencies (SWAs) in Nigeria when compared with the high production costs of water. Water rate in metered residential areas has been found to vary widely from one state to another. It is as low as NO.44/m^3 in Borno and Yobe States, slightly up to N2.6/m^3 in Kwara State, averaging N1.35/m^3 for the whole country. In the case of unmetered residential areas, the charge varies appreciably from N2.00/month in Edo State to N152.4/month in Anambra and Enugu States, averaging N32.5/month for the whole country. The production cost of water on the other hand is estimated at N20.00/m^3.

In a recent study conducted by the author, private water vendors operating in towns and cities in the northern part of the country were found to charge an average of N45/m^3, more than 33 time the tariff charged by SWAs for metered residential areas.

The paper examines the poor state of water supply in most towns and cities in Nigeria which can be traced to inadequate funding resulting from inability of SWAs to charge higher tariffs to maintain their services. Private water vendors who on the other hand are not licensed but free to fix their rates, make profits and expand their services to all towns and cities in the country. Other than the exorbitant rates charged by Private Water Vendors (PWVs), and that they sometimes obtain their supplies from unhygienic sources, nevertheless, their services go a long way towards supplementing government effort.

If the services rendered by the water vendors can be taken into consideration, there will be greater participation by the private sector in water resources development including their eventual involvement in the manufacturing of water equipment, plants and chemicals.

Background

Nigeria is blessed with abundant water resources. With an areal coverage of 924.000 square kilometers has a population by 1992 census of 88 million, which is about 1/5 (one fifth) of Africa's total population. An estimate of the available surface water resources of the country is put at 259 billion cubic metres, thus giving an average yield of about 283 litres/m^2. The underground water potential on the other hand is estimated at 87 billion cubic meters.

The country has varied climatic features as it extends over the tropical rain forest to as far north as the semi-arid zones. Despite abundance in water supplies, there is often uncertainty in the rainfall distribution pattern both in time and spatially. For example, the average annual rainfall amount varies from 3,000 mm/year in the southern part of the country to about on 500 mm/year in the extreme north. The annual amount varies widely from year to year, especially in the north and can be by as much as 100% from wet to drought year.

The Federal Government of Nigeria in an effort to improve the living standard and economic well being of the population has over the years embarked on many water projects aimed at self sufficiency in food production. The projects involve the creation of large capital outlay. Within a span of ten years, over 100 billion cubic metres of water had been brought under storage. Indeed, during the 1992 - 1993 period, water resources development has received more prominence than before, recording increase in budgetary allocation from 1.3% in 1992 to 3.2% in 1993 of the total capital expenditure. On the whole, the amount allocated to the agricultural sector, including water resources and rural development during the 1993 - 1995 Rolling Plan period add up to 14% of the total budgetary allocation.

History of water supply in Nigeria

The history of water supply in Nigeria at an organized level dates back to the colonial era when numerous water schemes were established to provide water to Government Reserve Areas (GRAs). The schemes were constructed and operated by Public Works Departments through levies or use of community labour force.

The system continued to independence whereupon, Water Supply Divisions were created to take over all aspects of water supply activities in general. The creation of states in the country in the early seventies, saw the emergence of State Water Agencies charged purely with the responsibility of providing water to both urban and semi-urban centres. In few cases however, the supplies were extended to rural areas as well. The SWAs are now in existence in all the 30 states of the Federation and in the Federal Capital Territory (FCT), Abuja, named either State Water Board or Public Utilities Board. In the latter case, water services are usually combined with electricity supply and road network developments in rural areas. These Agencies as created within the States are funded and controlled by their respective Governments through appropriate enabling edicts and acts.

At the state level, SWAs are generally charged with he responsibility for the development and supply of potable water to the inhabitants of the state. In general, they partake in the provision, operation and maintenance of urban and semi-urban water supplies. Current position show that about 80% of urban and semi urban population are served from groundwater, the remaining 20% from surface water. It is expected however, that the position would reverse in future due to excessive withdrawal from ground water sources and, more population would be served from surface water sources. Other governmental agencies, like Local Government Departments, Directorate of Food Roads and Rural Infrastructure (DFRRI)) and Agricultural Development Programmes (ADPs) also engage in water supply just as many non-governmental agencies like UNICEF, CUSO and WHO. There are also direct foreign government assisted programmes such as EEC, USAID, JICA, Chinese assistance, etc that engage mostly in borehole drilling and construction of hand-dug wells, particularly in areas to supplement water supply efforts.

Besides these bodies, there are private water vendors engaged in the development and supply of water to areas not accessible or connected to public supplies. The well established private vendors derive their sources from boreholes which they develop and supply water through tanker trucks and charge fees as they deem fit. The medium to small scale water vendors fetch water from stream sources or public standpipes, depending on the availability of supplies and their resources.

Problems of state water agencies

The State Water Agencies face quite a number of problems, mostly financial due to their total dependence on state subventions for both capital and revenue expenditures. This in turn subjects them to government regulations and often undue interventions in the day to day running of their organisations. A clear manifestation of this is that most SWAs are saddled with more office and administrative staff leaving the scientific service units under-staffed. Also political/social considerations are often brought to undermine their efficient performance and indeed, revenue generation.

Another problem recently identified[3] was, the high cost of water projects generally, resulting from the use of inappropriate design parameters. Prominent among these are, high per capital design allowances, use of large percentage of services through private house connections and, adoption of long-range planning horizons. These result in the over design of projects with excessive capacity and high capital costs, thus, leading to under-utilized investment and inflated operating costs. For example, the cost per capita for high level of service (90% house connections and 10% serviced by standposts). The present day overall development cost per capita, even assuming low level of service is estimated at N1,100.00 which given the prevailing worldwide economic hardships facing a developing country like Nigeria, is quite high. Identified also, is the poor financial performance of the SWAs. The low tariff they charge, is attributable to political considerations which inhibit the collection of even the little revenue due. It is worth noting that, in most of these agencies even if all possible revenues were collected, the cost recovery at prevailing tariffs would constitute only a small fraction of recurrent expenditures.

Role of private water vendors

Services provided by Private Vendors in developing countries are rarely appreciated. The low level of service provided by public water corporations, coupled with their inability to reach the remote parts of the cities, all tend to result in the inevitable involvement of PWVs in the development and distribution of water. The low tariff chargeable by the SWAs no doubt constitute a constraint to higher performance. For example, whereas, the present cost of producing one cubic metre of potable water, requiring pumping and chemical treatment is estimated to average N20.00, water charge for metered residential areas is as low as N0.44/m^3 in Borno and Yobe States, N2.64/m^3 in Kwara State, averaging N1.35/m^3 for the whole country[4]. For un-metered residential areas, water tariff rate is as low as N2/month in Edo State, reaching maximum of N153.4/month in Anambra and Enugu States, averaging N32.5/month for the whole country.

Private Water Vendors on the other hand, charge an average of N10.00/drum (220 litres) equivalent to N40/m^3. Alternatively, 16 litres tin which is popular with most residents is sold at N1.00, or N62.00/m^3. Water tanker trucks of 10,000 litres capacity sell between N300.00 and N400.00.

Private tankers usually source their supply from public utilities and are charged only N30.00/truck load for which the owners resell at ten times this cost. Assured steady profits generated by Private Water Vendors is the reason why they are expanding their services in cities and towns and getting organized.

In a detailed study (Whittington et al, 1983), of the role of water vendors in the supply of water to Onitsha town, in Anambra State, it was established that the vast majority of the residents obtain their water from well-organised water vending system, created and operated by the private sector. In this study, it was found that the Public Utilities supplied the residents only one third of their water needs while the remaining two thirds were supplied by water vendors. The revenue derived by the latter was also 24 times that by the former in the dry seasons.

The inevitable involvement of PWVs in the development and supply of water to cities and towns in Nigeria is quite evident; the choice for the government is to recognize the services rendered by them and legislated their existence. For proper accounting and better performance SWAs could also be given the power to regulate the activities of PWVs in order to protect the public from drinking polluted water as the latter would be made to conform to stated water standards, before they are licensed to operate.

Privatizing the SWAs would therefore be the only proper action that the government must take now, to arrest the eventual complete breakdown of the system in the near future.

Conclusion

Water resources development and supply in developing countries, particularly Africa, have been the exclusive preserve of government, through its agencies. With improved living standards and rapid population growth over the years both per capita and aggregate demand for water increased substantially, much beyond the capability of the agencies to sustain. These coupled with the global economic hardships, the existing systems, could not be properly maintained or expanded, thus, the situation led to the gradual deterioration of the facilities and hence, poor or inadequate services.

In Nigeria, the low level of service of State Water Agencies saw the emergence of Private Water Vendors, operating in varying degrees in cities and towns. They develop and sell water to public, charging fees as they deem fit and prosper, thus expanding their scope of services and getting more organized. The SWAs on the other hand due mainly to government policy as constrained by political and social considerations, for example, charge only 2% of the cost of producing one cubic metre of water. The low tariff chargeable by the SWAs is the main factor for their low performance. The government should licence Private Water Vendors and consider privatizing the SWAs as the only viable option towards promoting sustained water supply, now and in the future.

References

1. 1993 Budget Statement and the Rolling Plan, 1993-1995 Press Briefing by the Secretary of Finance, Mr Olashore O, 29th January, 1993, Abuja, Nigeria.
2. Dale Whittington, Donald T Lauria and Xinmin Mu: 'Paying for Urban Services - A study of water vending and willingness to pay for water in Onitsha, Nigeria'. The World Bank, *Infrastructure and Urban Development Department Report*. INU 40, March, 1993.
3. 'Federal Republic of Nigeria Water Supply and Sanitation Sector Memorandum', *World Bank Report No. 4696 UNI*, October, 1984.
4. Report on findings and recommendations on billing and collection system of State Water Agencies submitted by Snc-Lavalin Int. Inc. to Federal Ministry of Agriculture, Water Resources and Rural Development in Nigeria, December, 1992.
5. Nationwide Water Supply and Sanitation Rehabilitation Study: Financial Report submitted by Diyan Consultants in association with Binnie and Partners Consulting Engineers to Federal Ministry of Agriculture, Water Resources and Rural Development, Nigeria, 1988.

Utility development: Cairo, Eqypt

Dr Wilfred Owen Jr

DURING THE PAST DECADE, over US$3 billion were spent to rehabilitate and expand sanitary drainage and treatment in Cairo. This massive investment in new systems, facilities and equipment was a long-awaited opportunity to reorient and streamline the organizational structure of the Cairo wastewater utility and Ministerial policy units that regulate it. However, these sector reforms were not implemented. The construction of new works was the sole concern of the Government and donors.

With the commissioning of several new wastewater systems, the parties concerned shifted their attention dramatically to "protecting" the investments. They have relied upon short-term international operations and maintenance (O&M) service contracts. These costly contracts address performance problems specific to each pump station or treatment plant.

Ultimately, the Egyptian Government and donor effort to secure uninterrupted returns from the new infrastructure will depend upon reforms. A deliberate, long-term programme is required to establish the financial viability of the sector and to strengthen government policy institutions, which in turn influence the operational effectiveness of the utilities.

This paper describes the efforts of the Egyptian Government, donors and contractors to strengthen the institutional performance of the waste water utility during the construction and commissioning the new works. It describes specifically how training made a contribution to improved utility performance at a time when water/wastewater sector reforms had not yet begun.

The sector

The Cairo water and wastewater sector operated without a long-term, strategic plan during the past decade of public investment. A master plan for infrastructure development was produced but no sector-wide assessment was available until the World Bank prepared one last year. The attention of the Government and bi-lateral donors had been upon the visible need to prevent the flooding of streets and to remove wastewater from the city. Subsequently, the existing sewers (designed for 1.0 million people) were underlaid with three new systems able to convey and treat wastewater for more than ten of the city's sixteen million inhabitants.

There have been numerous reports by consultants highlighting the autonomy, cost-recovery and other institutional requirements for Cairo to receive sustained benefits from the sewer system investments. Since 1984, the Egyptian Government has signed agreements with donors accepting responsibility for O&M costs of the new wastewater removal, treatment and disposal systems. To date, the Government has not been able to cover O&M costs of the limited old system, much less that of the city-wide new systems. Macro level reform of the sector was not pursued. There has not been an increase in the real value of revenues collected and these are not yet retained by the utilities. Notwithstanding the O&M cost agreement, the Government still expects donors to fund O&M service contracts, thereby postponing again local institutional reform.

Technical assistance

Technical assistance attached directly to the Cairo wastewater utility began last year after the commissioning of the first new wastewater conveyance system. The institutional support consultant proposed several presidential and municipal laws. These would streamline the inter-agency requirements for autonomy, cost-recovery and would strengthen personnel and supply systems for effective O&M of the new sewer systems. The use of Egyptian company law was recommended to commercialize the Cairo wastewater utility. For the first time, efforts to strengthen the internal workings of the utility are being made in parallel with initiatives to address the macro issues of the sector.

The use of training

The utility and contractors used training during the rehabilitation of old facilities and the construction of new ones to upgrade the technical skills of the Cairo sewer system staff responsible for O&M. However, post construction services were slow to earn the same priority as construction supervision. Using engineers for training utility staff was viewed as untimely and having a high opportunity cost. It took manpower away from design, document preparation, site supervision, procurements, facility assessments, equipment repair, etc. Other constraints to the use of O&M training was the availability of competent instructors, tools and equipment. These were addressed one-by-one. The priority courses and on-job training activities were limited to specific equipment at specific new facilities. Each course was thoroughly prepared by an instructor team and was delivered in a timely fashion. Up front preparation was made also by the contractor and utility for the selection of a cadre of young trainees. They responded positively to the attention they received. Since training was a visible and easily-managed activity, it gained a reputation for being one for the more proficient utility and contractor activities. Course approval, preparation and trainee selection were decisions internal to the utility and therefore training was designed so that it could be implemented in-house. The annual training programs were set up in a way to avoid time-consuming interagency authorizations.

The resulting immediacy in the delivery of training was in sharp contrast to the lengthy efforts made to strengthen utility management, ie, the personnel and stores systems, the increase and retention of tariffs or the autonomy to the utility. During times when senior levels of Government were unresponsive to the reform proposals, momentum in management and technical training became identified as momentum in strengthening the utility.

Pre-commissioning period

During the pre-commissioning period, donors, contractors and the utility were concerned with the measurement of training, ie, trainee contact and classroom hours and total numbers of staff trained. While these criteria were easy to report, they did not promote the targeting of training to those work units with responsibility for priority tasks. Rather, the focus on quantities encouraged the offering of general or orientation courses given for short durations to diverse individuals throughout the utility.

However, once facilities and equipment were commissioned and operating under the full or partial responsibility of the utility, the quality of staffing and the training they received was viewed more critically. For a while, technical training was seen as a root cause of effective as well as ineffective performance of assigned staff on the new equipment. Eventually, supervision and safety skills and procedures were given due attention along with technical skills. In the end, training and O&M service contracting was accepted as one of many inputs to the successful operation and maintenance of a facility. Whatever the skill quality of the facility staff, be they contractor or utility technicians, it was always possible for their performance to be undetermined by poor management decisions at the utility headquarters or at the oversight Ministries, ie, the transfer of skilled personnel, underfinancing of maintenance budgets, biased procedures for the recruitment of new staff, incomplete delegation of responsibility to station supervisors, etc.

Post-commissioning training

After the new wastewater systems were commissioned in Cairo, the government water/wastewater policy and organizational structure of the utility surfaced rapidly as constraints to operations. Managers found themselves with wholly new equipment and systems, yet they had performance problems similar to those of the pre-World War II Cairo sewer system. Procedures for the supervision, promotion and remuneration and safety of personnel as well as the procurement, inventory and issuing systems for spare parts, tools and transport did not support work at the new facilities. Rather, they distorted performance. These problems stemmed directly from a lack of responsive management systems and autonomy. But unlike training, the control over these systems was located outside the authority of the utility. To influence them required a long gestation period and access to the highest ranks in government.

The fact that Government, donors and contractors delayed addressing these sector constraints until after the commissioning of the new investments resulted in a period of reliance upon costly international O&M service contractors. The contractors were requested to protect the new investments and to monitor facility operations for reasons other than their level of skill and experience. Rather, the contractors brought flexible management systems to Cairo that were used to by-pass the policies and procedures that constrained the local utility. O&M service contractors used foreign exchange to purchase and air freight spare parts. They stored, inventoried and provided immediate access to these spares for the work crews that needed them. Moreover, O&M contractors provided training and incentives to staff and their work crews operated on holidays and over weekends to maintain equipment and to attend to emergencies.

Recommendations

1. Investment in infrastructure should be authorized by Government and donors after guarantees are in place that ensure that the local utility is financially viable and able to take full responsibility for equipment O&M. Investments in utility facilities provide returns only when supportive policies, laws and institutions exist at the national level. The host-country ministries of economic planning or international co-operation and the program offices of donor agencies need to plan investments in a comprehensive way to ensure the feasibility of a flow of returns over the long run. In sum, investment is not equivalent to the completion of construction activities.

2. Contractors and utility leaders need to be aware of the power of the constraints placed upon their performance by sector-level bureaucratic systems. Responsibility for facility operations should be accepted by contractors when they understand fully the risks involved due to the constraints within the overall escort institutions.

3. Contractors can use training of utility staff as a visible, low-risk way to lean local day-to-day operating procedures as well as to provide competencies and confidence to the client. Appropriate skill training is a consequence of hands-on and classroom programs that are targeted to specific audiences as a series of interrelated training experiences, not as an individual course. Facility work groups and their supervisors are the key targets. The use of full-time, multi-disciplinary instructor teams that design as well as deliver training and technical assistance is recommended. Such teams must be built in a deliberate way over a period of time for them to act as models of inter-disciplinary understanding, ie, the civil, mechanical, electrical, safety, supervisory and financial aspects of facilities.

4. O&M and training specialists operating at the level of the facility need a direct exchange of information with top management of the utility and with others involved in sector reforms. In this way the lessons learned on site from applying new O&M skills and

upgrading supervision systems can be better used by management. The applications at the facilities provide immediate feedback to those guiding the reform effort and help estimate the real cost to operations due to sector constraints.

5. Reform in the institutional environment of the utility needs to take place prior to rather than parallel to internal efforts to augment skills and proficiencies. Long-lasting improvements require reforms in the water/wastewater sector that lead to a greater autonomy of utilities. Institutional development consultants should refrain from relying upon the easy-to-use tool of training to strengthen utility performance.

6. Training may attract priority because it can be a quick-starting activity, a deliverable and an action that visibly meets contract agreements. But the best training does not necessarily lead to performance improvements of a client. The acquisition of skills is only one of many factors contributing to performance. Project managers need to fight the "activity for activity's sake" mentality applied to training, especially when donors, contract officers and the utility apply pressure for "results." It becomes increasingly expedient to cite training activity as development activity.

In summary

Training can set up an exchange of views that leads to improved understanding and risk taking among utility, contractor and consultant staff. This understanding can be used to guide a reform effort while providing more time to deliver on hard problems. Training needs to be treated as a key component of the delicate process of reform. If executed well, training activities can expose which conditions precedent to facility and utility performance improvement are in place and can suggest which conditions are yet to be created.

Privatization of rural water supply

Mike Wood

SINCE INDEPENDENCE in the 1960s most governments of developing countries have been the main providers of improved rural water supplies and sanitation facilities. They adopted a socialist stance whereby the state provided for basic services. The rationale for this approach was that communities could not afford to develop water or sanitation systems themselves. At that time there were very few non-government organizations or private companies that could do such work. During the International Drinking Water Supply and Sanitation Decade (1980-90) of the total global funding for water supply and sanitation of $10 million, 65% came from state sources (1).

But over the last 30 years or so this approach has been only partially successful. Thousands of systems were constructed but at the start of the 1990s one in three people in developing countries still lacked access to the two most basic requirements for health and dignity; a safe and reliable water supply and adequate sanitation facilities (2). Although governments did manage to increase the number of people provided with water supply by 240% during the Decade, this does not mean that these systems are still functioning effectively today.

Because the emphasis in the past has been on constructing new systems, maintenance of these and older systems has been neglected. The figure of 40% of systems not working has often been quoted.

Background

During the 1970s and 80s the economies of most developing countries slid into decline as a result of falling prices for export commodities coupled with a dramatic rise in the cost of imports, particularly of oil. Consequently governments were obliged to borrow from international organizations in an attempt to balance their economies. These organizations forced governments to adopt harsh economic policies which caused a cut-back on social spending in the areas of health services and the provision of water supplies. This cut-back was felt more in rural areas which had less political clout than the burgeoning urban areas where population pressure forced governments to provide services.

Decline in government services

Some governments have also been spending a large proportion of their GNP on the military. For example in Ethiopia during the latter years of the Mengistu era, as much as 60% of GNP was being spent to fight insurgents in the north. This meant that the budgets for the provision of water and health services were cut back dramatically.

Government departments, strapped for cash, during the 1980s were unable to maintain the services they had provided.

During the 70s and 80s most governments' actively discouraged the private sector in favour of developing state enterprises. Hefty import tariffs and taxes on private property discouraged private initiative. In some Marxist oriented countries like Ethiopia, Mozambique and Angola, the state nationalized many private firms and confiscated their assets. Foreign investment was discouraged.

With the collapse of communism in the East and the rise in the influence of capitalism from the West, many developing countries have been forced to change their policies.

They have been forced to cut back their over-staffed bureaucracies. There have been massive lay-offs in the public sector. Up to 60 000 civil servants were made redundant in Ghana in the mid eighties, for example (3).

The emergence of the private sector

Countries adopting a market-oriented approach have been rewarded by support from international financial organizations and western donor governments. Ghana launched her own Economic Recovery Plan in the early 1980s which included measures designed to encourage foreign investment and to get private entrepreneurs involved in setting up small businesses.

Following the collapse of the Marxist regime in Ethiopia in 1991, the Transitional Government has adopted a similar economic recovery program. Private investors are now being encouraged by a number of practical measures which include:

- reduced import tax
- reduced level of income tax on private businesses
- a reduction in the amount of red tape required to get a business licence
- a devaluation of the currency, making imports cheaper (4)

In other African counties, however, the private sector is already well developed. In Kenya, for example, there are a number of private companies constructing water supplies and sanitation facilities.

In Burkina Faso, the Yatenga Project of the 1980s promoted the private sector in the maintenance of rural water supplies. Entrepreneurs were given assistance to purchase toolkits and received training from project staff. Spare parts for Village Level Operation and Maintenance (VLOM) handpumps were made available to local shopkeepers. Mechanics buy the parts and are paid by communities for repairing pumps.

In Ghana, the Wenchi Mission Water Project asked communities to pay 70 000 cedis (about $350 in 1986) into a rural bank account as a contribution towards constructing a borehole and installing a handpump. Some of the funds were set aside in a maintenance fund which is used when the Mission maintenance team is called to repair the pumps.

A water society in Kenya

There has always been a strong sense of self-help in the rural areas of Kenya expressed through the 'Harambee' concept. To fund the capital cost of a development project, money is contributed from each household according to how much it can afford. These funds are then augmented by the government.

In the early 1980s in Meru District on the lower slopes of Mount Kenya, seven small community groups existed with the aim of providing an improved water supply through gravity systems. In 1984 these groups decided to join together as individually they lacked a large enough membership to develop the extensive infrastructure required for piped water systems. They were also inexperienced in management and financial accounting and lacked the ability to mobilize their communities beyond initial fund-raising activities.

The groups amalgamated under the Murugi-Mugomango Water Society in 1983. The Society was constituted under the Kenya Societies Act, and so became eligible to receive NGO assistance.

The Ministry of Water Development helped the Society to approach NGOs for assistance in funding and for management training. The Canadian Hunger Foundation, an NGO, responded and contracted the services of Research and Planning Services, a Kenyan development consulting firm and Technoserve Inc., a Kenya-based American NGO. A trilateral agreement was signed between these agencies and Murugi-Mugomango Water Society. Technoserve and CHF assisted the Society to draw up a set of by-laws, rules and guidelines to govern the Society on a commercial basis.

This process was important in forging links between the self-help groups and helped to form a common consensus on the objectives and longer term goals of the Society. The NGOs also assisted the Society to enter into legally binding agreements with outside parties.

Laying the groundwork

During the first 18 months the Society achieved a great deal in setting up a workable administrative structure which reconciled the accounts and records of the former small groups.

Membership

To become a member of the Murugi-Mugomango Water Society, householders have to be approved by the Society's Management Committee. They then pay a registration fee of Ksh 4,170 (US$150 in 1991). Members are obliged to provide 60 days of labour for digging trenches and backfilling. Cash in lieu of labour is discouraged because it is felt that this labour contribution forms an important part of creating a sense of ownership of the Society and of the water system.

Institutions like schools and clinics pay Ksh.6843 (US$ 245) for membership. By 1991 the Society had a membership of 2,150 of whom 1,350 were connected to the water system. From membership fees the Society had assets of over Ksh.600 000 (US$21,500) and labour valued at over Ksh.1 million (US$35,700) had been contributed. In addition, members of the Society contributed Ksh.280,000 (US$10,000). The Society built offices and stores from 'harambee' contributions and from a Ksh.200,000 (US$7,000) grant from the government.

CHF provided over Ksh.2.8 million (US$100,000) in materials, technical assistance and training.

The gravity system

The Society was assisted by CHF and engineers from the Ministry of Water Development in designing the gravity system. Intakes have been constructed on the Meru River on the slopes of Mount Kenya. By 1991, over 380 km of pipeline has been constructed in main and branch lines along with six reservoirs and 20 break pressure tanks and valve chambers.

By mid 1991, 1350 consumers were receiving water through metered yard connections.

User fees

Each household pays a flat rate of Ksh.20 (US$ 0.75) for the first 30 cubic metres consumed per month. Consumers are then billed Ksh.2 for each additional cu.m. used. Institutions pay Ksh. 100 (US$ 3.60) for the first 30 cu.m. and Ksh. 3/cu.m above that amount. The number of defaulters has been very low.

Project management

The philosophy of the Society is to run it in a business-like manner so that the organization makes a modest profit. The Society elects a management committee at the annual general meeting. In 1991, eight members were elected and four, including two women, were co-opted. The chiefs and sub-chiefs of the area are among the members of this committee.

An executive committee of four of the top office holders is the Society's main decision making body. The Society employs 18 staff, headed by a project manager. Their salaries come from revenue generated from the tariffs which amounts to about Ksh 35,000 (US$1250) per month.

Problems

Although the Society is well managed, it is not without its problems. In 1991, the Society was only making a very small profit. This can be attributed to the comparatively low rate of the tariff, particularly for institutions. Commercial outlets like hotels are paying only Ksh. 100 (US$

3.60) for 30 cu.m. Hoteliers in the area agreed that they would be willing to pay much more for water. Many farmers use more than 30 cu.m. per month as they irrigate vegetables and coffee which they sell as cash crops. The levy of Ksh.2 per cu.m. appears low. However, the Society was unwilling to increase the tariff in light of low coffee prices in 1990-91. However, farmers were getting a good price for green beans during that time.

Water shortages have also been experienced during the dry season.

To alleviate this, the Society constructed a new intake and 7.8 km of pipeline to the main reservoir. The cost of this expansion was Ksh. 1.8 million (US$ 64,000).

The Society has been criticized for the low level of womens' involvement in the management of the organization. There were only two women on the management committee, and they had been co-opted. The reasons given were that members of the society believed that their interests would be better served by the men they had elected to the committee.

Observations

This Society provides a good example of a group of communities coming together in a sprit of self-help and being assisted in key areas of finance and training in which they were deficient, by outside agencies. These agencies have committed a considerable investment in this organization but they have been careful to allow representatives of the community to take major management decisions. Now the challenge for these external bodies is to gradually withdraw from the scene as the Society becomes more able to run its affairs in the best interests of its members in a more independent way.

This is just one example of privatization in the water sector. The following criteria can be drawn from this example which appear to be necessary for a private sector initiative to succeed:

- A community who need an improved system and who are prepared to commit their time and money to building it.
- A government who are prepared to back community self-help initiatives with cash and technical assistance, and which drafts legislation to encourage a business-like approach towards the provision of services.
- An external agency which is prepared to provide technical and financial assistance in return for accountability and a strong sense of community commitment.
- A community who have access to a cash economy.
- A community with leaders who are committed to the improvement of their community above their own self interest.

For privatization to work in the provision of essential services like water and sanitation, governments and external agencies must adjust their role to become more like facilitators for community initiatives and less like providers of systems for communities.

References

(1) *The Decade & Beyond*, Christmas & de Rooy, UNICEF, 1990.
(2) *The New Delhi Statement*, Sept.1990; UNDP.
(3) *West Africa Magazine*, Oct.1985.
(4) *Investment Code for Ethiopia*, T.G.E.,1992.

SECTION 5

IRRIGATION

Treated wastewater re-use in the Gaza Strip

Lana Abu-Hijleh

THE GAZA STRIP, occupied by Israel since 1967, is located along the Mediterranean Sea between Israel and Egypt. Its total area amounts to 365 km² and the present population is approximately 800,000 inhabitants, of which 50% are refugees.

The climate, which is typically Mediterranean, is characterized by daily average temperatures ranging from 13C° in January to 27C° in August.

The western part of the Strip is composed of sand dunes extending along the seashore, and the eastern part is composed of heavy soil of clay and sometimes of loose soil. The average annual rainfall increases from 200mm/yr in the south to about 400mm/yr in the north.

The total amount of rainfall is around 100 million cubic meters per year (mcm/yr), part of which is consumed by agricultural vegetation (45 mcm/yr) or is lost through surface run-off and evaporation (20 mcm/yr). The remainder infiltrates into the soil and recharges the groundwater (25 mcm/yr).

The major part of the water utilized in the Gaza Strip for domestic (25 mcm/yr), agricultural (100 mcm/yr) and industrial (5 mcm/yr) purposes is supplied from underground water. The deficit between the groundwater consumption (130 mcm/yr) and the groundwater recharge (estimated at 35 mcm/yr) has led to overexploitation of the aquifer, and has resulted in declining water levels and increased salinity reaching 1500 ppm of chloride in large parts of the area.

Short and long term solutions for the Gaza Strip water problem are being debated. They include:

- promotion of more effective water use in agriculture;
- protection of the groundwater quality by improving the sanitary situation and treating the wastewater;
- introduction of irrigation schemes with treated wastewater;
- collection of run-off rainwater for use in agriculture.

Other solutions which involve importing water from neighboring countries or desalination of sea water, in spite of being expensive, are being considered as possible long term approaches.

Within the framework of the short term action, from 1985 onwards, the United Nations Development Programme/Programme of Assistance to the Palestinian People has been implementing two major sewage collection, treatment and re-use for irrigation projects. The projects' objectives are to:

- reduce the use of fresh groundwater for agriculture;
- protect the groundwater from contamination by the infiltrating sewage;
- improve the environmental health conditions of the population;
- recharge the aquifer when possible with the treated sewage.

The first project was concerned with the upgrading of the existing sewage treatment facilities of Gaza City (with a population of 400,000 people) and the construction of an effluent distribution system to the citrus orchards south of the city. The main components were:

- Primary and secondary treatment lagoons (aerated);
- Gaseous chlorination station;
- Booster stations and storage reservoirs;
- Percolating ponds;
- Distribution pipelines to the citrus orchards and overflow line to the Gaza City Wadi.

The second project's intent was the implementation of the Sewage Master Plan for the Northern Region of the Gaza Strip (120,000 inhabitants). The region is composed of three towns and a large refugee camp, all closely located. When designed, the Master Plan intended to serve only the towns, ignoring the presence of the refugee camp despite its miserable environmental situation. However, following UNDP involvement in the project and by exerting pressure on the authorities, it was permitted to include the refugee camp in the master plan. The project components were:

- A complete sewage collection scheme for the refugee camp (70,000 people), considered to be one of the most highly condensed areas of the world;
- Lifting stations and a main pumping station;
- A central treatment plant (presently oxidation ponds designed to receive aerators);
- Distribution pipelines to citrus orchards and an overflow line to the Northern Region Wadi.

The completion of the two projects was delayed beyond the planned dates due to the political situation which prevailed in the Occupied Palestinian Territories (OPT) from 1987 onwards namely, the Uprising. The construction of the Gaza City project was completed in 1991 but it is not yet fully operational. The following describes the project's present status:

- Firstly, the effluent characteristics are not yet at the designed level, due to serious sludge problems in the ponds and the continuous breaking down of the aerators.

- Secondly, the chlorination station is not operational, because for security reasons the Israeli authorities do not permit using gaseous chlorine in the Gaza Strip.
- Thirdly, the farmers are still reluctant to irrigate with effluent and refuse to adapt their irrigation systems at their own expense.
- Finally, the overall management structure of the project is not yet identified, therefore; the municipality is trying to avoid the responsibility for operating and maintaining the project.

As for the Northern Region Master Plan, presently, around 80 percent of it has been accomplished. The following describes its status:

- The sewage collection component operates satisfactorily in the dry season; however, it floods and clogs in the rainy season because the population use it as storm drainage.
- The treatment segment also suffers from the flooding in the rainy season, which disturbs the whole treatment process.
- The irrigation part has had to be suspended until the farmers agree to irrigate with the effluent. The overflow line has met the same fate. Its construction was opposed by the land owners and the residents living adjacent to the Wadi, who fear the effluent will cause health hazards and odor problems in their area.

Due to the lack of relevant experience in this area and the fact that the UNDP was pioneering in undertaking projects of this sort and scale, most of the problems which have lead to the abovementioned situations were not anticipated and dealt with at the formulation or early implementation stages.

Only at the final implementation stages were the obstacles which hindered the full realization of the projects' objectives revealed.

The following is a personal contemplation of the multifaceted technical, institutional, cultural and political problems and circumstances which contributed in an integrated manner to the obstruction of the two projects. It is based on my experience as the projects engineer from 1986 to 1991 and the officer in charge from 1991 to date:

Technical

Most of the sewage schemes designs for the Gaza Strip municipalities are prepared by Israeli consultants. This is usually a condition set by the Israeli authorities in order for the municipalities to receive construction permits.

The designs of these two projects were also prepared by Israeli consultants and sponsored by the municipalities. UNDP, as the body responsible for the implementation of the projects encountered the following problems due to certain shortcomings in the designs:

a. The physical conditions of the Gaza Strip, such as the sandy unpaved roads and the lack of storm water collection schemes, were not taken into consideration. Consequently, the primary and secondary screening designed were inefficient, causing serious grit and sludge problems in the collection networks, at the pumping stations and at the treatment lagoons.

b. The second problem was related to the treatment system and to the equipment utilized. The designs were based on inappropriate technologies for the area: aerated lagoons, electro-mechanical grit removal devices..etc. Such sophistication and imported equipment require regular maintenance and highly skilled operators. In addition is the fact that some of the equipment was unfit for the Gaza climate.

c. Another shortcoming was related to the pre-design preparation and the construction implementation plans. It became apparent that the designers of the distribution and overflow lines did not investigate the ownership status of the land where the pipes were planned to be laid, i.e. public or private. Naturally, once the construction commenced, the private land owners objected and forbade the contractors to work in their territories. Therefore, the project was delayed tremendously; obliging the alteration of the designs. The delay, logically, has had an impact on the status of the whole construction plan. For example, the treatment plant for the northern region was constructed and operated prior to laying the overflow or distribution pipelines. Thus, the sewage was flowing in the ponds and had no way out except by flowing over the dikes into the adjacent inhabited areas, creating terrible health hazards.

d. Finally, the Sewage Master Plan of the northern region was limited in scope, covering only the population which resides in the three towns and excluding the refugees living in the camp (who comprise more than 50% of the overall population). This decision, I believe, was for political reasons, since the refugee camps are not recognized as permanent communities. However, once the refugee camp sewage collection network was installed and connected to the central treatment plant, the negative technical implications of such a decision appeared. The inflow to the plant was much higher than allowed for in the design criteria, disturbing the whole treatment process and necessitating significant changes in it.

Institutional

The capability of the counterpart institutions, namely the Gaza municipality and the Local Councils of the Northern Region, to take over and manage comprehensively the whole sewage schemes were not considered at the formulation or early implementation stages. It was only when the counterparts refused to take over the installations, that their incapability to assume the projects' responsibility was revealed.

At that time, and due to the fact that the newly constructed installations were deteriorating, the UNDP decided to launch another technical assistance project with the objective of enhancing the technical capabilities of the counterpart municipalities to maintain and operate the constructed sewage installations. For two years, and under the supervision of an expatriate and an experienced local engineer, a team of engineers and technicians employed by the counterpart institutions received the necessary theoretical and on-the-job training.

The project activities have been completed, and the team recently re-joined the municipalities to assume their responsibilities as the technical personnel in charge of the sewage installations.

This project dealt successfully with the technical shortcomings of the counterpart institutions. However, the lack of management capacity, combined with their financial problems, remains as a serious obstacle hindering the proper operation of the projects.

Another side of the institutional problems is related to the lack of a management structure to coordinate and organize the efforts of all the parties involved, i.e. the municipalities, the agricultural departments, the water departments and the farmers' representatives. For there were a number of complex, integrated issues related to the collection, treatment, or irrigation parts, which remained at another level difficult to be addressed by each institution individually.

Cultural

People's culture in all its aspects (religion, customs, traditions, ideals, perceptions..etc.) could act as a promoter or an inhibitor of a new idea. In the case of the sewage projects in OPT, specifically the treatment and re-use for irrigation part, cultural obstacles stood in the way of the people's acceptance of these projects as solutions to their water and environmental problems. I believe the following two points have led to such a negative attitude:

- To mix sewage with clean water, to work, or to irrigate edible crops with it, was perceived as an unholy practice by many of the farmers and sometimes workers due to the religious believes and customs. Such a view led them to reject any dealing with the projects, especially the treatment and irrigation parts.
- For the Palestinians to trust the good intention behind any project which concerns their community, it is extremely important for them to be totally involved in all its activities, including planning. Regrettably, this was not the case in these two projects. Consequently, once the projects were completed, the community members refused to assume any responsibility in operating and sustaining any part of the scheme, even the sewage collection one.

Political

In order to understand the connection between the political situation and the success or failure of the concerned projects, it is important to point out the following:

- In an attempt to keep an iron grip on the water resources in the OPT, the Israeli authorities do not release any significant data on the underground water quality or quantity. Moreover, they restrict the discharge of water from the wells owned by the Palestinians, by imposing certain quotas. These quotas are subject to reduction if the owners are proved to have alternative sources.
- As a result of the above, the Palestinians receive with suspicion any action which concerns water, even if it is claimed to be for their benefit. This was one of the reasons for the farmers' refusal to utilize the treated sewage for irrigation. For they feared it would allow the Israelis to reduce their fresh water quotas, and thus eliminate the historic rights they have to the underground water. This is a condition they will never accept under the present political situation.
- The Israeli Occupation imposes heavy taxation on the Palestinians residing in the Territories, claiming the revenue is invested in the development of the area. On the other hand, the Palestinian population believes that only a small fraction of the collected taxes are re-invested in the public service. For that reason, sometimes the community refuses to contribute or pay dues for any public service they receive, which they consider a right already paid for. This was the case in the sewage projects' operational cost: the communities refused to pay any dues for that purpose, therefore creating a lot of financial difficulty to the municipal structures whom the projects' maintenance and operation were entrusted with.

Conclusion and recommendations

The Gaza Strip water situation is presently considered to be at the crisis level. Scientists envisage a grimmer future in terms of the availability of drinking water for the ever-growing population in the Strip. Therefore, no possible solutions to this problem should be discounted, in spite of all the obstacles which may be encountered in implementing them.

Reducing the percentage of the fresh groundwater utilized for irrigation should continue to be the priority solution. Until now, the treated sewage remains as the most appropriate alternative for the fresh irrigation water, due to its economic feasibility and sustainment compared to other approaches.

This is the reason for the UNDP and the concerned Palestinians to maintain continuous efforts to overcome the obstacles facing sewage treatment and re-use projects. For their efforts to be fruitful, the right environment will have to be established and correct approaches adopted. This can be ensured, by drawing the appropriate lessons from the previous experiences encountered.

Within this context, the following recommendations can be made:

a. To involve the communities directly in all the project formulation stages: identification of needs, defining objectives, activities required, type and size of inputs, expected outputs, distributing roles and responsibilities..etc;
b. To integrate in any project framework an element concentrating on the enhancement of the counterpart institutional capacity;
c. To utilize Palestinian consultants in designing the required schemes and support them with all the technical know how needed;

d. To research and study the different alternative designs for sewage collection, conveyance, treatment and re-use, in order to define the most appropriate methods and technologies to be utilized in the Gaza Strip;
e. To launch a community awareness campaign which aims at raising the people's knowledge of environmental health issues, the importance of maintaining their sewage collection schemes and ways to do it, the water crises prevailing in the country and the vitality of treating sewage both for environmental and economic purposes;
f. To launch another specialized educational campaign targeting the farmers and covering the following: Their contribution to the water crisis, the importance of modifying their present irrigation schemes, the viability of treated sewage as an alternative for irrigating with fresh water, the methods of treatment and all related monitoring control and health precautions required;
g. To construct a demonstration plant for treatment and irrigation. Such a plant could be utilized for the research and the public education purposes;
h. To encourage the formation of a central sewage authority for the Northern Region of the Gaza Strip including the Gaza City. Such authority would comprehesively manage all the sewage schemes within the designated region.

References

Dr. H.J. Bruins, Ir. A. Tuinhof, Ir. R. Keller (1991) *Water in the Gaza Strip* final report to the Government of the Netherlands.

United Nations (1991) *Israeli Land and Water Practices and Policies in the Occupied Palestinian and other Arab Territories*.

United Nations (1992) *Water Resources of the Occupied Palestinian Territories*.

Trickle irrigation using porous clay pots

S K Agodzo, J W Gowing, and M A Adey

TRICKLE IRRIGATION CAN result in very high water use efficiency and is well-adapted to cropping on marginal land but is not an appropriate technology of the majority of small-scale, resource-poor farmers. Simple, low-cost applications of the trickle concept have received little scientific attention. This paper presents summary of an extensive laboratory, field and computer study into such a system based upon the use of porous clay pots.

Background

Pot irrigation is thought to have originated in North Africa (Barth, 1988) and provides a simple and efficient alternative to the highly sophisticated and expensive forms of trickle irrigation.

The general principle involves embedding unglazed earthenware vessels in the ground, to be periodically filled with water and covered with an earthenware lid (Barrow, 1987). Alternatively, a number of porous capsules may be automatically filled by a network of interconnecting pipes under pressure (Silva, 1980). The hydraulic permeability of the pot means that water flows into the soil surrounding the vessel, thus providing a moisture supply to the root zone.

Given that there is continuity through the wall of the vessel, then the rate of water release depends upon the potential difference between pot and soil. Therefore, the system is, in principle, able to respond automatically to changes in the rate of root water uptake. The nature of this response is, at least in part, dependent upon the properties of the ceramic material.

Experimental method

In this study, the behaviour of the pot-soil-plant continuum was investigated through

- measurement of hydraulic properties of ceramic pots,
- field experiments over two seasons in Ghana, and
- development of a computer model of the pot-soil-plant system.

The experimental methods are briefly explained below but full details can be obtained from Agodzo (1993).

Hydraulic properties of ceramic pot

The object of this investigation was to establish the manufacturing conditions ideal for clay pots used as irrigation supply units by examining the roles of firing temperature and clay texture in influencing the hydraulic properties of the pot.

Three general purpose smooth, medium and coarse clays were investigated. Using samples of the pot wall the porosity was determined by weighing of vacuum-saturated samples and the water retention curve measured using pressure plate apparatus. Smaller cylindrical pots were manufactured and the saturated hydraulic conductivity measured using standard falling-head permeameter procedures.

Field experiments

The field experiments were designed to study the pot-soil-plant continuum and to evaluate the effects of pot sizes and plant population on plant water use and yield. The experimental site (Kumasi, Ghana) experiences a sub-humid tropical climate with average annual rainfall of 1300 mm and potential evaporation of 1260 mm. The field experiment comprised four blocks each of six randomly assigned plots (4 m x 7 m). The treatments were factorial combinations of three water treatments and two planting densities. Watering treatments consisted of a control (rainfed) and two sizes of pot (3 and 4.5 L respectively). There were either one or two plants/pot. Two test crops were used, pepper (Capiscum frutescens) grown in the dry season (1 November 1990 - 28 February 1991) and okra (Hibiscus esculentus) grown in the wet season (8 May - 14 August 1991). Plants (with their associated pot) were grown on a 1 m x 1 m grid.

Results and discussion

Hydraulic properties changed both with clay texture and firing temperature (Agodzo et al. 1990). Coarse clay resulted in the largest porosities and smooth clay the lowest. As firing temperature increased from 850°C to 1250°C the porosities decreased from 26% to 15% in the case of the coarse clay through to 21% to 0% in the case of the smooth clay. Water retention curves indicate that, for all three clays, the majority of pores have an equivalent cylindrical radius < 0.3mm. However, with increasing firing temperature pores < 0.3 mm are lost and the proportion of pores between 0.3 and 1 mm increases with the largest such increase occurring for the coarse clay.

Hydraulic conductivities of material fired at 850°C are similar ranging from 0.1 to 0.3 mm/d. They increase with increase in firing temperature to a maximum in the range 950°C to 1150° and then fall. This increase is much the greatest for the coarse clay which rises to 1.8 mm/d.

In using the pot as an irrigation device, there is the need to thoroughly investigate the hydraulic properties of the pot since the malfunction of the pot as a water application device may result in crop failure. From the investigation it was concluded that a temperature range of 950°C - 1050°C is required for the production of the irrigation pot, corresponding to 18% to 24% porosity and hydraulic conductivity 1.4 mm/day is achievable using coarse clay. Limited investigation of traditional clay pot firing techniques in villages around Kumasi indicated that firing temperatures seldom exceed 750°C and consequently hydraulic conductivity values are likely to be relatively low.

Pots were filled on a weekly basis with demand varying between 1.4 to 2.5 l /week. The field experimental results showed the pot regulating itself to maintain a balance between water supply and demand (figure 1). Plant density per pot was the dominant factor in overall fruit yield, given that other growth factors were adequate. Pot size did not affect yield, but irrigated treatments showed a clear improvement over the rainfed control.

Under irrigation, commercial yields of pepper are in the range of 1.10 to 1.25 t/ha fresh fruit (Tindall, 1983) for a minimum economic life of 6 months. Pepper yields for the irrigated trials lasting 4 months ranged between 1.0 to 1.3 t/ha. According to Tindall (1983), commercial okra yields can be up to 2-3 t/ha. Up to 2.0 t/ha of okra were realised for the irrigated treatments.

Maximum irrigation effectiveness of 1.73 kg/m^3 and 3.86 kg/m^3 were obtained for pepper and okra respectively. Irrigation was more effective during the dry season than the wet season. No previous figures have been reported in the literature for the irrigation effectiveness of pepper and okra.

Laboratory-based lysimeter studies indicated a zone of influence extending radially some 10 cm beyond the perimeter of the pot. Field observations indicated that the majority of roots grew as a sheath immediately around the pot. Clearly the pot distorts the normal root growth habit. It is thought that root suction may control pot outflow directly.

Conclusion

Production methods clearly influence the hydraulic properties of the ceramic material used. In particular, firing temperatures may be crucial in determining the resulting hydraulic conductivity.

Under the conditions of these field experiments, the volume of irrigation water supplied by the pots was small, amounting to approximately 50% of the pot volume per week. Nevertheless, the pot supply was responsive to demand and this was important as reflected in the increased yields.

Rooting habit is strongly perturbed by the presence of a pot. This places a premium on consistent pot maintenance as the soil water reserves will be less readily available to the rooting system. While no problems apparently arose in these experiments, it might also mean that, under more arid conditions, water supply to the roots might be limiting. This is being explored using a mathematical model of the pot-soil-plant system.

Acknowledgements

The financial assistance and support of both GTZ and DAAD is gratefully acknowledged.

References

Agodzo S K (1993). *Trickle irrigation using porous clay pot*. PhD thesis, University of Newcastle upon Tyne, UK. 292pp.

Agodzo S K; Gowing J W; Adey M A (1991). 'Use of porous clay pots as alternative trickle irrigation device'. In *Techniques for Environmentally Sound Water Resources Development*, R Wooldridge (Ed), Pentech press. 10pp.

Barrow C (1987). *Water resources and agricultural development in the tropics*. Longman Dev Studies, London.

Barth S (1988). *Irrigation by earthenware vessels - an alternative for arid and semi-arid zones*. GATE Publication, Bonn, Germany, 3, 26-27.

Silva D A (1980). *Irricao por capsula porosa. Caracteristicas e evaliaco do metodo sob pressas hidrostatica*. Master's thesis, Universidade Federal da paraiba, Campina Grande, Pb, Brazil, 67p.

Tindall H D (1983). *Vegetables in the tropics*. Macmillan Press, London.

SECTION 6

SANITATION

Makata pumpable VIP latrine block

A H Abel and S V Dohrman

THE MAKATA STYLE pumpable VIP Latrine Block is a block of twelve latrine stalls and four urinals combined in one structure. Each stall has its own concrete squatting slab, its own pit, and its own vent pipe with attached flyscreen.

This latrine block was designed specifically for a crowded urban setting where a lot of latrines need to be provided. A standard septic tank pumper truck or another method of removing the excreta out of the two meter deep pit must be available if this style is to be used.

An extensive survey of Blantyre, Malawi's existing primary school latrines and toilets was done while the Makata Latrine Block was being designed. Lessons learned from that survey shaped the design of this latrine especially in the many small but significant design details.

Some of the reasons this system might be chosen over other designs are:

- This system isolates all waste matter *including* the water used for cleaning the latrine which is often neglected in other block designs.
- This design saves space. This feature is especially desirable in urban situations where schools are overcrowded and land is needed for future building extensions or playing fields.
- Residents of urban areas find the single attractive building more acceptable than standard individual latrines which may be seen as a step backward.
- The Makata block latrine uses strong, basic materials and calls for a very high quality finish. It is therefore very long lasting and requires little maintenance.
- The relatively shallow pit design can be built in an area with poor soils where a deeper excavation would normally need to be braced and shuttered.

Other features of this latrine are:

- Extremely low maintenance costs.
- Pumpable with normal septic tank pumper trucks.
- Uses basic materials that can be found in most local hardware stores.
- Low operating costs consisting only of pumping costs and the water used for washing the floors. The pits need to be pumped out about every two years. Each pit has a capacity of 5.8 cubic meters.
- Uses modular pre-cast concrete blocks to make the vent pipes.
- Having facilities all in one place makes cleaning easy.
- Virtually theft proof.
- The material cost is only about 35% higher on average than an equivalent number of individual latrines but the Makata style latrine can be expected to last far longer.

Major problems to avoid

The design *details* of this system set the Makata style latrine apart from more basic designs. These details were developed from our experience in the field. Before designing this latrine every toilet and latrine in the City of Blantyre's Primary School System was inspected to get an idea of what worked and what didn't work in the existing facilities. These details were developed to avoid the problems that were discovered in this survey.

Standing water inside of buildings

This problem is usually caused by poor design. Most facilities in the survey lack floor drainage or are poorly constructed. The result is that low spots in the floors become puddles of filthy water that can remain for days. This water is a problem because diseases breed and live in it. They can then be passed on to the many people who come walking through the facility.

Muddy, wet areas outside the entrances and exits to latrines and toilets

This situation occurs if the designer fails to put an apron, sidewalk or drain at the doorways to the latrine or toilet facilities. Wet, muddy places pose the same problem as standing water.

Foul water draining into areas nearby the latrine

Since there are often no floor drains or soakaways provided to collect fouled cleaning water, that water is, by necessity, swept out the building only to find its way into a nearby road, path or residence. This poses health threats for the students as well as people who pass by or live nearby. The problem can be avoided by using a soakaway.

Stalls and corridors that are too large

Stalls and corridors that are too large encourage children to defecate indiscriminately in those areas.

Excessive darkness and poor ventilation

The problems mentioned above of children defecating everywhere except for in the toilet or latrine is worse in facilities that are too dark. Young children who are sometimes afraid of dark places and are often the worst offenders. They defecate in the corridor rather than risk going into the dark stall. Adding breeze blocks as windows in each stall and large open metal gates at the two entry ways brightens this design. This increased light has not in anyway increased fly breeding. These improvements have the added benefit of allowing more air into the latrine which helps to dry the floor out

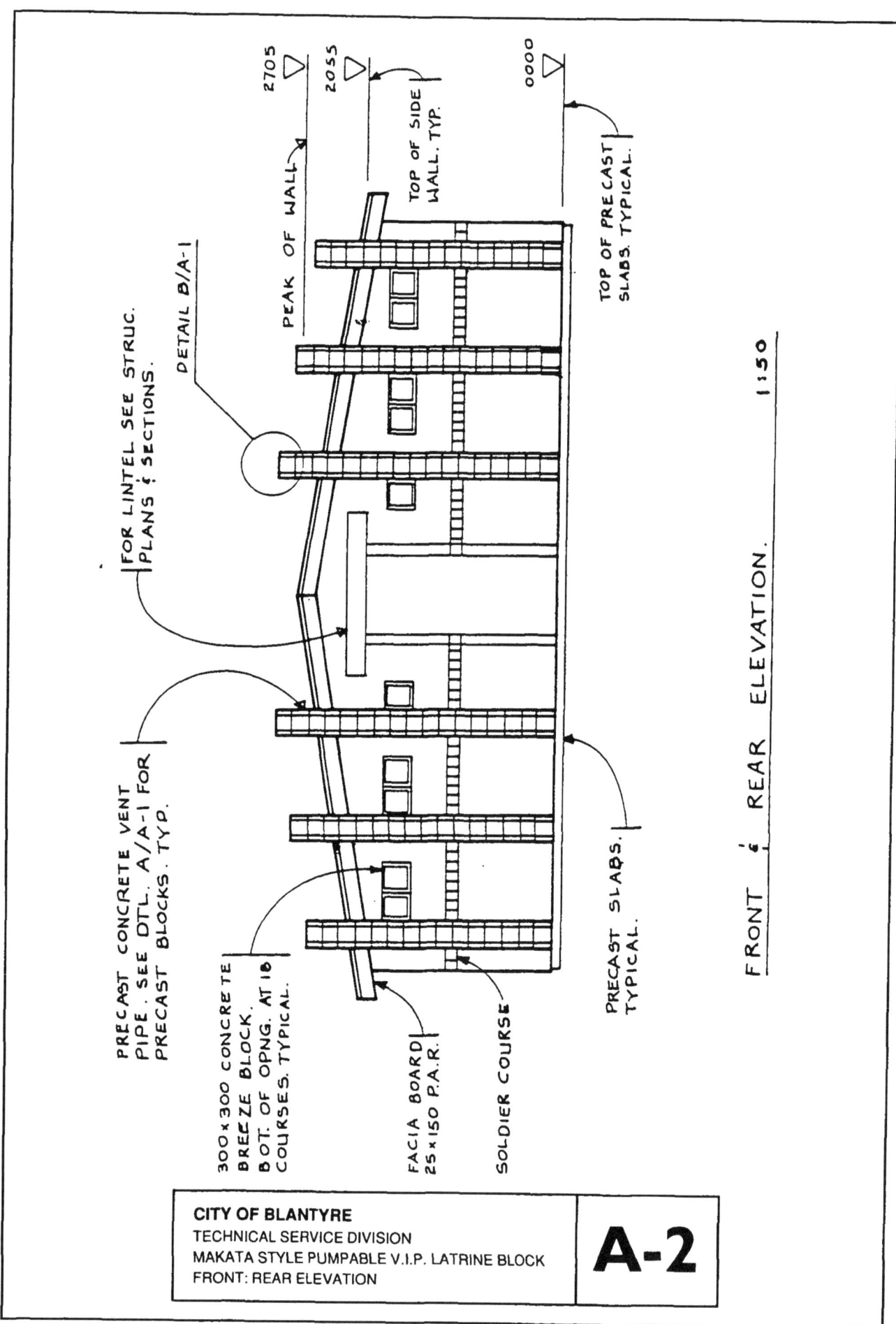

SANITATION: ABEL and DOHRMAN

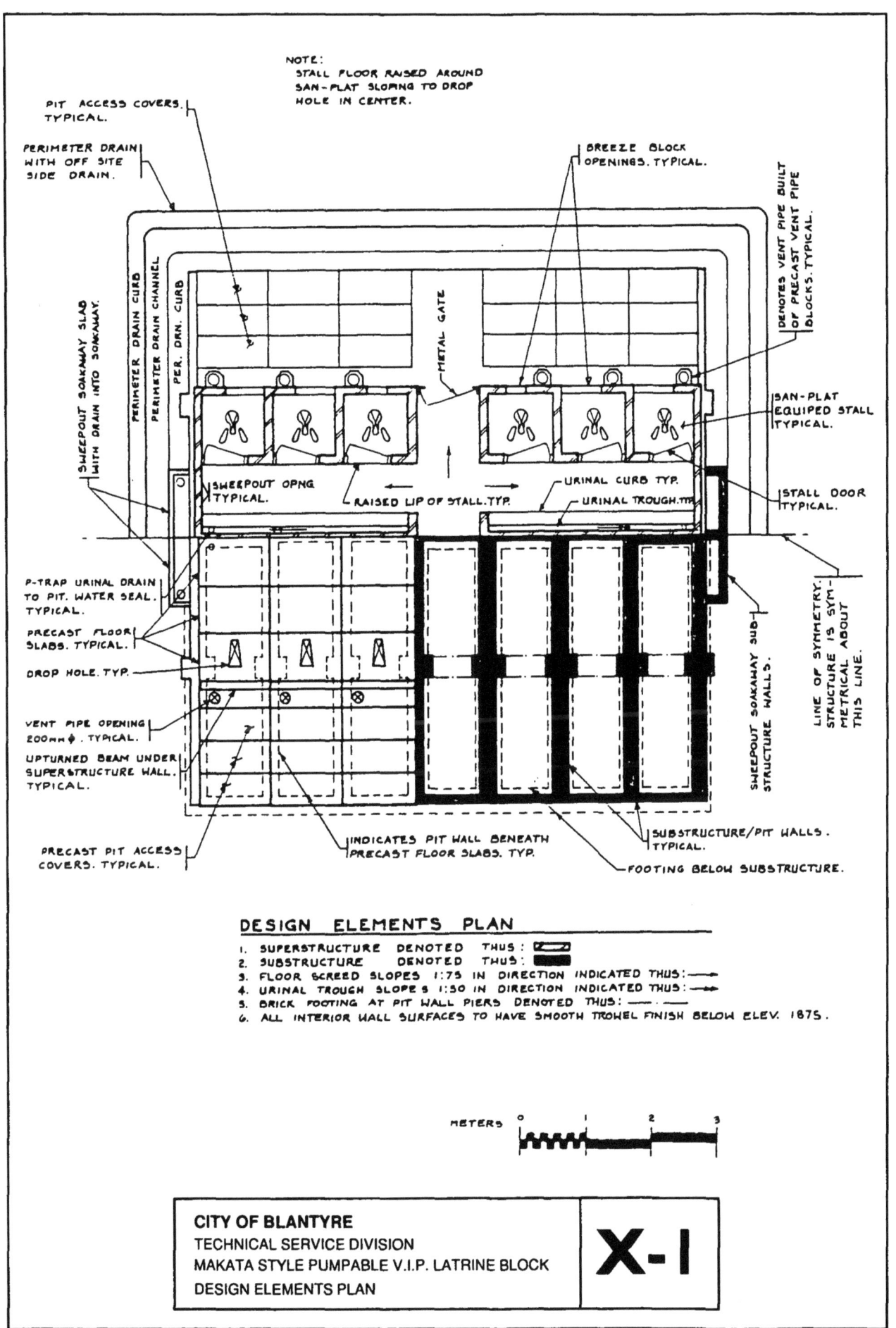

CITY OF BLANTYRE
TECHNICAL SERVICE DIVISION
MAKATA STYLE PUMPABLE V.I.P. LATRINE BLOCK
DESIGN ELEMENTS PLAN

X-1

between uses, killing off disease vectors that live only if water is present. Toilet and latrine floors should dry out completely at least once a day.

Design details

The following design details were conceived to prevent the problems listed above:

One pit, one vent pipe, one drophole

This main principle behind the Makata latrine's design is the same as the basic principle of the standard VIP Latrine design. For every squatting slab there must be a vent pipe and a separate pit. In other block latrines where one large pit is used for several squatting slabs the latrine doesn't vent well and so the smell can be quite offensive. In this design there are twelve separate latrine stalls and four urinals. Therefore, for one latrine block there are twelve squatting slabs and twelve vent pipes in the superstructure and twelve separate pits in the substructure.

Sweep-out opening and soakaway for cleaning water

At the end of the corridor there is an opening through the side wall at the floor level. The water used to clean the floor is swept out the opening. Many designs make no provision for getting rid of this water and so the cleaner sweeps it out the entrance or exit creating a dangerous mix of water and disease where everyone is sure to step. Care is taken to insure that the floor in the corridor is well sloped. This insures that all of the water in the corridor leaves through the sweep-out and there is no chance of water standing inside.

Next, a soakaway has been provided adjacent to the building to safely isolate this water. The soakaway is quite narrow in order to prevent rainwater from the roof from pouring into it. The cleaning water runs through the building wall and immediately into this soakaway. This feature solves both the problem of standing water inside the building and of water running onto adjacent areas.

Exterior drainage

Exterior drains all around the building shed the water that collects from the roof, and intercepts rainwater that may flow toward the building. The drains also shed any water that is used to clean the inside or outside of the latrine and is not intercepted by the soakaway.

Small stall and corridor sizes

In the pre-design survey it was found that if the stall was too big, larger than one square meter, children would often not use the drophole but instead would defecate in the open space or in the corners. The smaller sized stall has worked really well to solve this particular problem. For the corridors the same problem exists. When the corridor is too wide or if there is a large open area, which is common with waterborne toilets, children are often found urinating or defecating in these places. By decreasing the corridor size, getting rid of unnecessary area and by eliminating blank walls from the boys latrine by providing urinals instead, this problem has been solved. By making the corridor smaller the traffic in the latrine effectively makes it impossible to misuse the corridor. It is easier to use the facility correctly than to misuse it.

Squatting slabs

The squatting slab is a standard feature in most VIP Latrines. It is a small, hard concrete slab with a keyhole drophole and strategically placed, raised footrests. It is easy to clean as the floor slopes toward the drophole. A lid is not recommended in this application.

Further information

If you want plans or other information about the Makata Latrine write to: *Stewart V Dohrman, 3604 Trail Ridge Rd., Louisville Ky 40241, USA* or *Alexander Abel, 310 Lookout View Ct., Golden Colorado 80401, USA* or the *San Centre Supervisor, Post Bag 67, Blantyre, Malawi.*

Rural school sanitation pilot project

Rob Burgess

IT WAS IDENTIFIED in Umgeni Water's Rural Areas Water and Sanitation Plan (RAWSP), that sanitation at almost all rural schools is inadequate or missing entirely. This situation has undoubtedly arisen to some extent because the schools do not receive any subsidy from the relevant authorities for sanitation facilities.

Statistics from 1991 revealed that a total of 417 schools existed in the RAWSP study area. This represented an enrolment in excess of 200 000 pupils and a capital investment in excess of $6 million will be required to provide adequate sanitation.

Background

The motivation for targeting schools is to ensure that the younger generation have the opportunity to learn about and experience healthy and safe sanitation. This information should then be carried home with them to the benefit of all. There is also the strong motivation that water and sanitation should be implemented simultaneously. The health benefits are self evident, but the administration and cost recovery procedures can then also be linked to complement each other.

To date the implementation of water schemes, as identified in RAWSP, are on programme, but progress on sanitation has been limited. This is caused by the fact that water is seen as a priority need amongst rural communities, whereas sanitation is not in such high demand. Umgeni Water is generally reactive in implementing water schemes, whereas sanitation will require a proactive approach.

Pilot project

In order to obtain more insight on how to overcome these problems it was decided to implement a pilot project in Sweetwaters, a peri-urban area 20km outside the city of Pietermaritzburg. Umgeni Water had recently completed construction of a water reticulation scheme in the area, which included water supply to seven schools. These schools ranged from Junior Primary to Senior High schools with enrolments ranging from 600 to 1300 pupils.

Initial contact with these schools revealed their keen interest in the project and also the urgent need for improved sanitation. There was no official school maintenance service in the area and this responsibility was carried out by the school staff and parents. The area was served by gravel roads in poor condition and no telephones, electricity supply or sewer reticulation was available. Local building contractors were available but had limited skills and would need assistance if employed on the projects.

Steering group

A steering group was set up by Umgeni Water to address various aspects of the project. These included community/school liaison, appropriate technology selection, pollution control, funding and administration. Represented on this group were the Institute of Natural Resources of the University of Natal and Umgeni Water. As progress was made the School Circuit Inspectorate and School Health Inspectorate were consulted for their input.

Appropriate technology selection

This group initially debated the basic criteria required for selection of the suitable technology. The facts taken into consideration were that the system should comply with the required health standards, have limited water consumption, minimal maintenance requirements and the construction method should be labour intensive and cost effective. The remoteness of the area and the limited availability of services also influenced these decisions.

After various considerations and extensive community liaison it was decided to implement a non-flush dry-septic system with various improvements, as recommended by the steering group. The septic tank was constructed using a "Riblock" system of precast concrete ribs, interlinked with concrete blocks. The inside of the tank was plastered and a suitable waterproofing compound applied. The toilet blocks situated adjacent to the central septic tank were built using "Lassak" precast concrete products. The wall panels hung vertically from a ring beam supported on uprights. The base of the panels were then tied in with the floor slab. Ventilation and natural lighting was provided by the opening between the sloping fibre-cement roofing and the ring beam.

The vertical toilet pedestal and horizontal pipe section connecting to the septic tank were manufactured as one solid unit out of fibre-glass. During the construction of the tank these were concreted in position as they accessed the tank through the side walls. Thus when construction of the toilet block took place, the pedestals were already in position. To avoid blockages, the diameter of the vertical section of the chute below the seat was kept to a fairly wide dimension. The chute then turned through a ninety degree angle. The horizontal section was designed with an increasing diameter towards the tank to ensure it would be blockage free.

Percolation test/soak-away

Umgeni Water Scientific Services staff carried out soil percolation tests to assess on-site soil conditions and size the soak-away. Due to the relatively low inflow of a dry-septic system the siting of the soak-aways proved relatively simple and could be accommodated within the school property. To enhance the performance of the soak-away a lining of coarse gravel was used against the invert and sides of the trench and broken clay bricks were used as backfill material.

Community liaison

In parallel with selecting the appropriate technology, liaison was taking place with various parties. The School Staff, School Parents Committee, and School Inspectorate were involved at all stages of decision making. The proposed offer of a "soft loan" from Umgeni Water was accepted by the Parent's Committee and then the proposed project was presented to a full parent's meeting. A favourable vote indicated their approval of the project and signalled that construction could commence.

Labour resources

The school Parent's Committee recommended a list of names of unemployed parents available for employment on site. After interviews with these people, all positions other than site supervisor were filled and a standby list was prepared.

As was anticipated, an initial high degree of close supervision was required on site. The building method was unfamiliar to the labour but transfer of skills took place rapidly. Efficiency on site was motivated by the employees having a sense of ownership of the project.

As work progressed a core-group of labourers were selected and it was agreed that they remain with the project to carry their skills forward. It was also envisaged that a multiplier effect would be possible as this core-group gained experience and could then be divided up and allocated to various projects simultaneously.

Commissioning

Before commissioning of the toilets took place, the local Health Inspector informed the staff about the correct use and maintenance of the toilets. They in turn passed this information onto the pupils. At this stage the problem of toilet paper was raised. The staff did not feel the school could afford this item. It was initially of concern that the use of anything other than normal toilet paper would upset the functioning of the septic tank.

Various discussions and calculations regarding anticipated tank retention time followed. A retention time in excess of 21 days was agreed upon, therefore the septic tank performed as a digester and would be able to handle heavier types of paper. The pupils were in the habit of using old exercise books in their existing pit latrines and it was decided to let this habit continue.

Problems

Soon after commissioning of the toilets, alarm was expressed by the staff that the toilets were blocked. It then became apparent that their perception was that these toilets would be full flush. Observation also revealed that the contents of the chute did seem to float for an unexpectedly long duration.

This raised another problem. A type of blue bottle fly (chrysomyia putoria) was breeding in the contents of the toilet, the eggs were incubating and the larvae managed to wriggle up the side of the chute and onto the floor.

This caused great concern to all involved. Investigation followed where the larvae were identified but no immediate solution to the problem was forthcoming. Any form of insecticide used would possibly poison the tank. The larvae were extremely sticky and able to crawl over any type of surface. Introducing mechanical means of preventing the flies access to the chute would result in a possible increase in the maintenance requirements.

The breeding of larvae rose to alarming proportions almost causing the toilets to be abandoned. It was then observed that the contents of the toilets was sinking into the tank at an increased rate. In conjunction with this the appearance of the larvae started to diminish. It was then deduced that possibly with the tank becoming more activated the process of digestion had increased and the required egg incubation period had been removed from the breeding cycle.

Improvement in financing scheme

During this time separate negotiations had been underway, the result of which was that a 50% grant funding was now available to the schools for sanitation. Initially a piped water supply had been excluded from the project as funding and control of consumption were of concern. The schools now expressed their interest in a water supply in the toilets in order to assist cleaning purposes and hopefully enhance the digestion process in the chute. The staff were prepared a supervise cleaning operations in order to control consumption. It was then concluded that water would be supplied to the hand-basins in the staff toilets. For cleaning purposes a hose-pipe was supplied to wash down the rest of the toilet block and break-up/saturate the contents of the chute.

Lessons learnt

Other than the unexpected appearance of the larvae the project presents a variable solution to rural school sanitation. With a few examples now in operation and available for site visits no false expectations should be raised by future interested parties.

The way forward

To simplify construction procedures and reduce costs changes to the design are under consideration. The recommendation is that the toilet blocks will move to directly above the tank, sharing a common back wall along the centre of the tank and a central rainwater gutter. The roof slab of the septic tank will form part of the floor slab of the toilet block. This will simplify the chute design and septic tank water proofing requirements as the chutes will now penetrate vertically through the roof slab and extend to below the water level in the tank. Observation of the commissioned toilets revealed the necessity for urinals to be provided for the boys. This could be achieved, quite simply, by redesigning the top of the chute in the form of a urinal.

Reference

The Institute of Natural Resources, University of Natal. Umgeni Water's Rural Areas Water and Sanitation Plan (RAWSP)

Subsidy: to what extent?

Shamsul Huda

RURAL SANITATION IS one of the Government of India's (GOI) important programmes in the current five year plan (1992-97). Compared to its Rural Water Supply Programme, which carries a very high political commitment, Rural Sanitation was a late starter, though, it was included under the "20 Point Programme" in 1985-86. Centrally Sponsored Rural Sanitation Programme (CRSP) has been in operation since then. As it happened in many developing countries, in India too, the CRSP was launched with a high subsidy based service delivery.

Challenge

Can India with a rural population of 627 million people (1991) afford to provide subsidy for its Rural Sanitation Programme? Let us consider the achievement status at the end of 1990, at which point the government estimates a 3% coverage in Rural Sanitation. If this trend continues then achieving global goal i.e. **universal access to sanitary means of excreta disposal by 2000** is going to be a major challenge for India. Fund requirement to reach the Global Goal for Rural Sanitation is estimated to be Rs.300 billion (US $ 10 billion) which is far beyond any proximity. The eventual questions are "Will it be possible to provide this basic facility to the families?" "How long will it require?" "What about the improvement of quality of life of rural habitants?" And so on. However, a survey conducted by the National Sample Survey (NSS 1988-89) has revealed that around 11 % households have latrine facilities. The incremental 8% (excess over 3% coverage through government programme) can therefore, be attributed to private sector and can be considered as a spread effect of the government programme. The private sector thus possesses more potential in terms of providing services to the community for adopting sanitation practices. This has been a eye opener for the policy makers.

Alternate approaches

UNICEF provides support to the Rural Water Supply & Sanitation programme in India. Although in financial terms this support is very low (2-3% of the total outlay) its strategy has been very useful in bringing qualitative improvement in programme delivery. Demonstration projects undertaken with UNICEF collaboration has shown a distinct difference in terms of cost effectiveness and enhanced community participation through alternate approaches and has indicated a fair possibility for wider coverage. Alternate delivery systems, package concept of sanitation and appropriate technology promotion with a back up support of Information Education Communication & Social Mobilisation have been the main features of these demonstration projects. The experiences thus gathered can be categorized in to three major groups ie:

a) Latrines could be constructed with nominal subsidy,
b) NGOs can promote sanitation without any subsidy for construction and
c) Sanitation promotion is commercially viable.

Eventually there has been a little shift from the conventional approach and methodology for implementing the sanitation programme.

IERT initiative

Under the GOI Rural Sanitation programme the two pit pour flush latrine is being promoted. In case of Uttar Pradesh state the estimated construction cost of such latrine is about Rs. 1500 (excluding roof and door of the superstructure) of which Rs. 1200 has been subsidized by the government while the beneficiary family pays Rs. 300. As a shift from this approach Institute of Engineering and Rural Technology (IERT), Allahabad demonstrated sanitation promotion through reduced subsidy at the rate of Rs. 450 per latrine. To achieve this the IERT through its community polytechnic system utilized the voluntary work force (Health Workers) in its extension areas. While the sub-structure of the latrine followed a standard design, the selection of superstructure was left with the beneficiary families. Of course a range of superstructure designs including their cost implication, labor intensiveness in construction and also the limitations were explained to the interested families. The beneficiary families, based on their best judgement, selected the design for the superstructure. In most of the cases the family members themselves constructed the respective superstructure. On completion of the construction work they were reimbursed with the subsidy component.

CAT initiative

Centre for Appropriate Technology (CAT) in Kanyakumari district of Tamil Nadu also demonstrated a similar approach. The CAT is a voluntary organization whose major work is in the fishing villages, majority of the habitants of which belong to the economically weaker section. CAT promoted pour flush off set single pit design through its extension workers. In this case the estimated construction cost is Rs. 2000, against which a subsidy of only Rs. 500 was provided. The interested families were provided with the design, possible option of construction materials for both sub structure and superstructure and also with a list of trained mason. The families either utilized the mason directly or they entrusted CAT to undertake the construction work. In this case too, the subsidy component was reimbursed after the construction work was complete. In some cases CAT also arranged bank loans for the families, payable in instalment with soft interest.

RKM initiative

In Medinipur, West Bengal, a sanitation project was implemented through an NGO, Rama Krishna Mission (RKM) which followed a different approach altogether. RKM utilized the network of village youth clubs and women voluntary groups for implementing this strategy. The uniqueness of the project is the full cost sharing by the families for construction of sanitation facilities. Medinipur project provides variety of technological options (10 designs of latrines with cost ranging between Rs. 230 - Rs.3000) to suit the different socio-economic segments of population. These options must provide water seal pans of any kind which ensure maintaining sanitary quality. A sound IEC backup and an in-built credit facility for construction attracted even the poor farmers to have their own latrines. A recent quick survey conducted in 17 villages of 8 blocks in Medinipur shows that about 79% of the families opted for the designs in the cost range Rs.230 - 280, while 2% opted for designs in the cost range of Rs.500 - 700, leaving 19 % who opted for designs in the cost range Rs.750 - 3000. Survey also reveals that more than 60% of the families who constructed their latrines fall below poverty line. The lesson learnt therefore is, that if a motivated community is provided with the right technological option then financial capabilities are not a barrier to sanitation promotion.

RSM initiative

Rural Sanitary Mart (RSM) is another approach to develop an alternate delivery system to facilitate improved rural sanitation coverage. The RSM is a retail outlet dealing not only with the materials required for construction of latrines and other sanitary facilities but also with those which are required as a part of the Sanitation Package. Besides, the RSM also serves as a counselling centre for creating demand amongst the potential buyers. For this purpose, it maintains a stock of models and manuals on sanitation facilities. Thus the RSM is, in a way, a service center too. Uttar Pradesh has been the first state to start establishing RSMs in 1991, through Panchayat Udyog (cooperative society) which is a Rural Industrial Complex promoted by a group of Gram Panchayats (Village Council). The state has now 12 such Kendras located at selected places. While all these units are engaged in manufacturing mosaic pan/trap and pit cover, some of them also produce items like cattle trough, fuel-efficient chulha, food safe, foot wear, water drum, storage bins, roof and door for latrines, hand-molded fibre glass pan and trap, and so on. RSM receives a support @ 25% of yearly turnover as seed capital. In addition, a managerial support for 2 years and lump sum support for marketing are also provided.

Experiences of an RSM at Bakkas (Lucknow district) show that promotional costs for latrine construction is as less as Rs. 50 per unit (family) which in comparison to the total business made, appeared to be commercially viable. Based on the feedback received from Uttar Pradesh RSMs have been established in Rajasthan, Orissa and Delhi. Government of India already issued a memorandum requesting State Governments to try RSM approach. With further strengthening the communication and social mobilisation components, RSMs possess a great potential for promoting sanitation and safe water and these also may establish linkages with control of diarrhoeal diseases.

Opportunity

These experiments have largely influenced in redressal of the whole sector by the government. One main achievement has been that recently the government has decided not to provide subsidy for latrine construction for the population segment who are above poverty line. The present thinking amongst the policy makers is to under play the role of subsidy in the promotion of sanitation. Rather, sanitation is to be taken as a people's movement to be adopted in their way of life.

References

1) *A Programme for Children and Women in India, Plan of Operations, 1991-1995*

2) *Eighth Five Year Plan, 1992-97*, Government of India.

3) *National Seminar on Rural Sanitation; Problems, Propects and Strategy for Future*, Govt. of India, New Delhi, Sept 1992.

4) *Water Supply and Sanitation*, UNICEF Donor Reports, 1991 and 1992.

Sanitation and solid waste disposal in Malindi

Dr Joyce Malombe

LACK OF ADEQUATE sanitation and solid waste disposal facilities is one of the major problems facing secondary towns in Kenya. The paper briefly analyses the situation in Malindi town. The data used in this paper is obtained from a preliminary analysis of data from an on going research project on sanitation in Malindi town. The Malindi Township is situated on the Kenyan coast of the Indian Ocean and lies in Malindi division which is in Kilifi District. The town is located about 120 kilometres north of Mombasa. The Municipality covers 334 square kilometres of land and 230 square kilometres of the Indian Ocean. The town of Malindi is characterised by a well developed tourist industry and subsequently a big portion of working migrants from several regions of Kenya contributing to the heterogeneous composition of the local population. The estimated population in 1992 was 57,000 inhabitants (Republic of Kenya 1992: 1-10). The Malindi population is composed of a mosaic of different groups differentiated by ethnicity, language, religion, income level, and occupation. Like in other secondary towns, the Municipality is characterised by a high rate of low income households. According to 1992 estimates, a total of 61 percent of all the households belonged to low-income group while 27 percent belonged to middle-income group. Only 12 percent of all the households were estimated to be in the high-income category. The level of income is one of the major determinants of the location of the settlements. For example, the closer the beach, the better-off is the population. The poorer income groups are found to the west and southwest where the living conditions are very poor. Most of these households work in different sectors. A survey done in 1991 (Gauff, 1991) estimated that 55 percent of the working population is self-employed while 45 percent is employed in the formal sector.

The high rate of migrant labour to the town has resulted in lack of basic services. In addition development of housing and other related services have been affected by different things. One of the major problems is the existing land tenure system in Malindi town. The land tenure system in the Municipality is still characterised by the historical ownership pattern of Arab absentee landlords many of whom still live in the Arab Peninsula. Absenteeism hampers the identification of owners, the survey of plots as well as the collection of land taxes (rates). Land ownership patterns include traditional Arab owners of residential, commercial, agricultural or idle land. European owners (or seasonal residents on the plot), European owner who runs a hotel or lodge on their plots, and government and municipal land. In addition there are households who reside in Malindi or elsewhere who own a house but not the plot where it is located, paying an annual rent to the landlord or a middleman. At times residents are tenants who are renting both land and house or only land or only the house and others who are simply squatting.

These land ownership patterns hamper development of infrastructure because of inadequate land available to the Municipality and the fact that tenants cannot make decisions on improvement of services because they do not own the plot and cannot decide on whether or not new latrines are to be constructed.

Service provision

Service provision is one of the major challenges to secondary towns and Malindi is no exception. The survey indicated that the town lacked many services and efforts to provide services did not reach many households. One of the major services needed is water. The survey found that although a piped water network exists in Malindi town it is inadequate and many households experienced water shortage. Other households could not afford to connect water to their houses and these bought water from their neighbours. It is worth noting that there is acute water shortage because most of the taps are usually dry and this makes water unnecessarily expensive. Landlords have also dug their own wells however, the water in some of these wells is contaminated and this is a health hazard.

Sanitation

Malindi town is not equipped with a central sewer and most people either use septic tanks or have pit latrines. High income households and the hotels have septic tanks. These tanks are however, at times too small and in poor condition. In general, most of the households use pit latrines. These are mainly situated in the middle and low income housing areas as well as in rural or mixed (urban) areas. There are very few latrines in the high income areas and these are mainly for domestic servants. The low income households in particular use pit latrines which are provided by the landlords. Not all the households have latrines and a total of 24 percent of the households in Malindi did not have any form of sanitation. The remaining 76 percent had some type of sanitation but most of them needed upgrading.

Four types of latrines were found in Malindi town and these included traditional latrines with wooden slabs and wind shelter, improved pit latrines with masoned pits and solid compartment, Ventilated Improved Pit (VIP) latrines and pour flush. Most of the latrines were found in residential areas and only a few were found in the industrial sector and public institutions and in commercial sector. The total number of traditional latrines has been estimated to be roughly 1,590, whereas the improved latrines were roughly 760 (Republic of Kenya, 1992). When industrial and public sector are considered the total number of pit latrines in Malindi is estimated to be 2,470. It was observed that often the traditional latrines do not meet the required hygienic standards. Some of the traditional latrines were also inside the house and smell was a nuisance in most cases.

Septic tanks and latrines are usually emptied but emptying and disposal services are inadequate. There is one official service and 7 private companies dealing with emptying of septic tanks and pit latrines and all these have limited capacity. In addition the services are too expensive for the low income groups. This is a big problem in plots where only tenants live but not the owners. The tendency is to leave the latrines overflowing for some time until the tenants take action by reporting the case to the Municipal Health Department. It is also not unusual to find overflowing septic tanks in both people's homes and hotels. These apart from being a health hazard have very bad smells.

Dumping sites are in two locations and these are situated in abandoned quarries. No environmental protection measures have been taken to avoid possible underground pollution which could come from infiltration of excreta liquid, heavy black oil from the cotton seed processing plant, used oils, grease from garages and service stations and soluble toxic substances not included in solid waste. These sites therefore constituted significant environmental risks. Lack of consideration of environmental problems caused by improper dumping of waste is a big problem in many urban areas in Kenya.

It is recommended that priority should be given to the construction of latrines for about 24 percent of the resident population without any sanitation facilities and mostly belonging to the low-income group. Another 30 percent of the resident population is still using traditional latrines which need an improvement to VIP latrines or even pour-flush latrines. In addition, another 13 percent of improved pit latrines, VIPs or pour-flush latrines need to be upgraded. These figures show that 67 percent of Malindi residents either do not have sanitation facilities or have inadequate sanitation facilities. Households with adequate sanitation only represent one-third of the total urban population in Malindi. This represents a small population with adequate services and given the high rate of growth of this town, lack of sanitation services is likely to bring about many health problems to the residents. This is especially true given the fact that the households lacking these services are in crowded low-income neighbourhoods.

Families have traditionally provided themselves with latrines and do not usually wait for the authorities to provide these services. The main problem is adequate finance to construct better latrines. It is recommended that low income neighbourhoods be given some assistance to construct better latrines and this can be recovered from tenants who complained bitterly about the standard of latrines provided. It is also recommended that a central sewer system be provided for the densely populated areas of town and the hotels. Most of the cost of construction of the sewer can be recovered from the hotels and rich households living near the beaches. The environmental problems caused by dumping of waste from latrines and septic tanks should be assessed and dealt with before it is too late. Better dumping sites should be provided and companies dumping their wastes in these dumping sites should be charged. The Municipality should not wait until it is too late to correct the situation. Households should be encouraged to construct better latrines, using appropriate technology which would reduce the number of times that the latrines have to be emptied. Construction of traditional toilets without lining should be prohibited given the fact that many households are still using water from the wells most of which is said to be contaminated. In addition, the contaminated wells should be identified and use of water from these sources discouraged.

Solid waste disposal

Like in most secondary towns, solid waste collection and disposal is a major problem in Malindi town. The increasing population in the town has added to this problem. For example, in 1991 it was estimated that 36,000 tons of solid waste were produced in Malindi, whereas only 7,300 tons were transported to the dumping sites by the Municipality collection services. Refuse collection is only done by the Municipality who have 5 tractors and two other vehicles. These are inadequate for the amount of garbage produced in the town.

Solid waste is normally stored within one's compound. The refuse is stored in dust bins or drums supplied by the Municipal Council. There is an acute shortage of supply of these containers by the Municipal Council and where such containers are not provided concrete slab containers have been constructed for storing refuse. In some areas refuse is dumped in one corner of the compound awaiting collection. A big portion of the solid waste is not collected but is scattered around the settlements. Many of the bins can be seen overflowing throughout the settlements.

The collection of refuse by the producer is done within the compound. Due to irregular service rendered to the producers by the Municipal Council, efforts are made by producers to find ways of disposing off the refuse. The main methods of disposing off the refuse include burning, composting or dumping of refuse anywhere. At present there is very little recycling and composting being done.

There is only one dumping site managed by the Council. But there is no proper system of dumping, it is done haphazardly, wherever space is found. The dumping site does however not meet the minimum requirements with respect to technical provisions like bottom and lateral sealing. The Municipal Council does not even have one bulldozer at its disposal to compact the dumped wastes and to cover them with soil.

The main problems faced include insufficient and inadequate storage bins, vehicles and management problems. The main constraints in the existing solid waste management is the lack of adequate staff, and equipment. The drums supplied by the Council, if any, became very heavy, once full, and cause a problem when being lifted by refuse collectors. Each vehicle is accompanied by four collectors who often have to empty the drums onto open plastic sheets and then lift these sheets onto the loaders or trailers, making the whole process very slow and loosing finer material. No proper maintenance of the vehicle is carried out at the Council yard. For any major repairs the vehicles have to be taken to Mombasa hence causing delays in the operation. The dumping site is becoming a health hazard to the public in the surrounding areas. Smell, smoke from the burning refuse, flies and mosquitoes are now becoming a nuisance.

Conclusions and recommendations

The above discussion shows that Malindi has poorly developed infrastructure system and communication network. Squatting, uncontrolled development of housing settlements impede long term planning and development of sanitation facilities. In addition complex land tenure structure affects decision-making for any new developments.

To deal with the above problems it is recommended that the water quality of wells be investigated and further pollution stopped. Those wells already contaminated should be put out of operation. Public toilets should be inspected, maintained and repaired. The Municipality should impose treatment of the by-product of black oil on a cotton seed oil company at the company's expense. In addition, the Municipality should promote and support private enterprises with respect to cesspit emptying and refuse collection services. They should start introducing standard bins with lid for domestic and commercial refuse storage facilities, purchase a bulldozer for the refuse dumping site to compact waste immediately and to cover them regularly with soil and clean the existing drains.

References

REPUBLIC OF KENYA, Ministry of Local Government and Municipal Council of Malindi. 1992. *Malindi Sanitation and Hygiene Education Feasibility Study: Inception Report*. Vol. 1 & 2.

GAUFF CONSULTING ENGINEERS. 1991. *Malindi Pipeline Project – Malindi Consumer Survey and Block Mapping*.

MANICIPAL COUNCIL OF MALINDI. 1988. *Local Authority Development Programme*.

An alternative pit latrine emptying system

Maria Muller, Jasper Kirango, and Jaap Rynsburger

THIS PAPER ADDRESSES the development of an appropriate pit emptying service, including the design of suitable equipment, in Dar es Salaam, Tanzania. The basic perspectives which guided the project partners are presented as well as some information on how the Manual Pit Latrine Emptying Technology (MAPET) service is functioning. MAPET is community based, but will provide better service if integrated in the city-wide service system of Dar es Salaam. Project partners for this pilot project (1988 - 1992) were WASTE Consultants and the Dar es Salaam Sewerage and Sanitation Department.

Situation in Dar es Salaam

In Dar es Salaam, as in other large Third World cities, the great majority of houses have on-site sewage disposal, i.e. mostly pit latrines, some septic tanks. Pit latrines are used by 80% of the households. On the 1992 population of over two million inhabitants or 450,000 households, this means that Dar es Salaam has about 170,000 pit latrines. Obviously, when the pits are full, they must be emptied[1]. It is estimated that yearly about 50,000 m³ of sludge from latrine pits need to be emptied. Add to this the demand for the desludging of septic tanks, and one realises that any pit emptying service agency faces a formidable task. Are the authorities in Dar es Salaam able to respond to this demand?

The Dar es Salaam City Council operates, through the Dar es Salaam Sewerage and Sanitation Department (DSSD) and the Health Department, its own vacuum tanker services with about five cesspit tankers in continuous operation each.

Apart from the formal system, there are informal, self-employed, pit emptiers who practise the traditional method[2]. Characteristic of this method is that, next to the full latrine pit a shallow hole is dug on the resident's plot, and that the sludge is scooped into this new hole by manual labour. Another characteristic is that, the pit emptier and the house owner deal with each other personally, without the interference of a (bureaucratic) organisation. In a process of face to face negotiations they agree on the price to be paid and the day of starting the work, and on the location of the hole for burying the sludge.

The existing services together do not have sufficient capacity to handle the rising need for pit emptying. A major shortcoming is that the voluminous size and weight of the vacuum tankers is unsuitable for narrow and unpaved roads in the densely built, unplanned areas. Especially the low-income areas lack adequate services because of the unsuitability of the vacuum tankers. The main requirement was, therefore, to design equipment appropriate for the densely settled areas; equipment that is manufactured and maintained locally. However, technical innovation alone is not enough to improve service delivery.

An alternative service

The new equipment and service is called MAPET (Manual Pit Emptying Technology). DSSD took responsibility for introducing MAPET through its own organization in Dar es Salaam, while WASTE Consultants acted as the advisor. The equipment is manually operated and is sufficiently small to be manoeuvred through narrow roads. Using local materials and components and widely known construction techniques, the equipment can be locally produced and repaired in small workshops. The operation of the equipment requires team work of three men, who - as experience bears out - stay voluntarily together for several years. As MAPET can function to a large extent independently from a centralized administrative organization and workshop, it is possible to decentralize its service to the neighbourhood level.

MAPET technical features and operation

A MAPET team consists of three men. One is the leader. In order to be allowed to rent the MAPET equipment he needs a certificate from DSSD. For this certificate the team must first do a training at DSSD. If a pit emptier is found dumping the sludge somewhere behind the bushes, he loses his certificate.

The MAPET team goes with two hand carts (one pump cart and a tank cart of 80 cm width) from the community centre to the customer. They can cover a distance of a couple of kilometres. They first negotiate with the customer where to dig a hole to bury the sludge. They then insert the hose-pipe into the squatting hole and connect it to the tank cart. The tank cart is connected to the hand pump with an air hose. The air is pumped out of the tank and the resulting vacuum causes the sludge to be sucked into the tank. The full tank is emptied into the hole.

Digging the hole constitutes most of the work and takes more than one hour. The 200 litre tank is full within five minutes. With heavy sludge it takes longer. Water is mixed into the sludge. By draining the hose-pipe out at full vacuum ('plug and gulp') the sucking can be intensified. Customers generally ask for 4 to 10 tanks to be taken out of their latrines. The pit emptiers earn about 2,000 to 5,000 shilling which they share among themselves. In order to make a living of the MAPET pit emptyings they should have at least one customer per day.

The process of MAPET introduction

The following points of view have guided the development of MAPET:

First, pit emptying is a service consisting of several components, of which the equipment is only one element. Other components are e.g. training to operate the equipment, repair facilities, the capacity to find customers, economic and financial aspects of the service organisation, and facilities for sludge disposal. All these components of the MAPET service have subsequently been addressed during the pilot project. Project experience has confirmed the importance of appropriate and locally constructed equipment. It has also confirmed the notion that a service can only be performed satisfactorily if all other components function properly.

Secondly, the introduction of new equipment, even more so of a whole new service, requires a step-by-step approach. This allows the innovations to be adjusted to local conditions at the appropriate time. This entailed e.g. that the basic MAPET equipment was constructed as a proto-type in a few months' time, but that serious adjustments were made in response to the experiences of the immediate users, i.e. the pit emptiers, over a period of 3 years. Similarly, training of the mechanics took place over a number of years, as they carried out the improvements in the MAPET equipment in the DSSD's own workshop. A step-by-step approach also implied that other components of the MAPET service were developed only when the need arose. For example, when the pit emptiers found it difficult to generate a regular demand from customers, a system of informing and motivating customers and community leaders was developed.

Thirdly, the new service, including the equipment, should be based on the most appropriate elements of the existing pit emptying methods. That is, building upon what exists, on what is known and familiar to people and organisations. In this way MAPET is not a strange element, as it combines e.g. the modern vacuum technology of the cesspit tankers with the traditional system of onsite sludge disposal by manual labour. It also strengthens the so-called traditional element of personal interaction between pit emptiers and customers, which is an important feature of modern small-scale, informal business contacts.

Fourthly, a form of public-private cooperation was envisaged between the DSSD and the informal sector. The public authorities have ultimate responsibility for sanitation services as they concern public health. It was also recognised that the demand for employment is tremendous. In times of structural adjustment programmes, MAPET could not generate new employment opportunities in DSSD, a government institution, but only in the private, informal sector. The solution adopted was that the DSSD would be the owner of the MAPET equipment and lease it to the pit emptiers. The DSSD provides essential support services, such as performing large repairs, promotion of MAPET in new neighbourhoods, and training and supervision, while the pit emptiers are self-employed workers, responsible for earning their own income. They do not receive a basic salary from DSSD. In this cooperation DSSD has a position to control irregular sludge disposal by private emptiers.

Different forms of organisation and management are conceivable, with a different balance between public and private responsibilities. Several options are being tried out in Tanzania.

The resulting MAPET service has both advantages and disadvantages. Some of the advantages are that:

- The MAPET equipment can reach the most inaccessible houses.
- The service can be performed almost immediately, while the vacuum tanker service requires a long waiting time.
- And the possibility of regular social contact between residents and emptiers, which enables community influence and supervision.
- MAPET can offer 'service to size': small volumes suiting the customer's household budget.

Some of the disadvantages are that:

- The MAPET service is expensive per unit of volume (m^3) compared with that of the vacuum tankers.
- The method of sludge disposal (burying on the plot) is not suitable for areas with a high ground water table and very densely populated areas.
- Cash flows between the DSSD and the private pit emptiers are difficult to control in practice.

MAPET service as part of a city-wide system

The pilot project has shown that MAPET can function satisfactorily in local communities. The emptiers can identify their customers and earn a low but steady income, informal mechanics in the neighbourhood carry out minor repairs, a certain amount of sludge disposal takes place within the community, and in a general sense MAPET enjoys social acceptance in those communities where it is already working. Leaders in other areas that came to know about MAPET are eager to bring it into their neighbourhood as a solution to the public health problems. Some NGO community initiatives have identified MAPET as a first priority to start a neighbourhood improvement campaign. On the other hand, residents and leaders would like to have more influence on the MAPET service, as they observe the potential for integration within the economic and health service system of the local community. Also they see the potential for income generation by the community.

However, MAPET is not an independent alternative to the tanker service. The size of the population requires the volume and hauling capacity of pit emptying as performed by the DSSD vacuum tankers[3]. In addition, MAPET should be operationally linked to the DSSD regarding sludge disposal. In areas with a high ground water table, MAPET cannot operate at present because of the absence of disposal facilities. Sludge must be removed from these areas and transported to central dumping stations of the city. The DSSD is the most likely organisation to use its vacuum tankers for this purpose. The aim is to combine the advantages of a community based service with the advantages of a strong organisation able to haul sludge through the city for final disposal. The required institutional arrangements (technical, financial, and operational) between the DSSD as a bureaucratic, government controlled organisation, the independently operating MAPET pit emptiers, and local communities are quite complicated. This is a formidable task, not less than the first introduction of MAPET.

The next phase of the MAPET project will include the development of an institutional framework for a neighbourhood based service, as well as the development of a sludge transfer system. The sludge transfer will initially be directed towards locally manufactured transfer stations as well as options for sludge treatment at neighbourhood level.

As in the first stage of the project, progress will be directed by the problems experienced by the organizations and operators directly involved at the city-wide and at the neighbourhood level. Solutions will be reached through a unique combination of the potential of these organizations in the public, private and community sectors.

References

1   *Comparative Study on Pit Emptying Technologies*, Draft Final Report, WASTE Consultants, Gouda, The Netherlands, 1993

2   *MAPET Progress Report 2*, WASTE Consultants, Gouda, The Netherlands, 1988

3   The COMPET study has recommended to separate urban areas with pit latrines into typical large tanker, mini tanker and MAPET areas. The typical MAPET areas are those where even mini tankers do not have access. Large tankers appear to be the most economic (if adequately managed, which is often not the case) for hauling sludge to sludge disposal stations over distances more than 5 km from the pit.

Low-cost sanitation and GARNET

John Pickford

ONE OF THE difficulties in trying to investigate water and sanitation practice in developing countries is that field workers are usually too busy to write about what they have done. When they do find time to set pen to paper (or fingers to a keyboard) it is often to commend something that they have done or plan to do. Consequently the literature has many accounts of projects and methods that are claimed to have been successful.

There is also plenty of advice, for example on systems that *ought* to work or ways in which communities *ought* to be involved. I well remember in the early 1970s meeting someone who was then a young engineer, full of youthful enthusiasm for a particular technique. He then set out to prove the value of his method. For twenty years he has tried to show that his idea was a good one. At least he should be congratulated for persistence.

What is lacking is *objective* reviews compiled by those who do not have a cause to proclaim nor axes to grind. It is difficult for the people who are actually involved in programmes to describe their own failures, and anyway the 'actors' in projects often move on, leaving behind the supposed beneficiaries who have to live with the failures. But we need to know about methods that do not work if we are to make progress with systems that are likely to be successful.

One research project did seriously attempt to compare alternative sanitation systems. It was carried out by the World Bank in the late 1970s in preparation for the International Drinking Water Supply and Sanitation Decade. Field studies including cost estimates were carried out in several countries by Bank consultants. There followed a series of publications (Kalbermatten et al, 1981) which were widely distributed and have been so constantly referred to since that everyone who has done anything about sanitation knows them. However, in spite of the tremendous effort put into field studies (and the undoubted very high cost since this was a World Bank job), the research was far from world-wide. The data was from very few countries. This was probably an inevitable result of employing paid consultants. In retrospect it can also be seen that the Bank studies and the conclusions drawn from them reflected the economic optimism of the time. There are many mentions of upgrading progression starting with VIP and pour-flush latrines.

Yet there is evidence that the greatest need is less for upgrading sequences than for *lower cost* sanitation methods. My present research seeks to ascertain amongst other things the extent to which cost of sanitation affects coverage. I am trying to look at low cost sanitation thoroughly and objectively. I want to find out what really works in practice and, more importantly, what stays working - what is sustainable.

I have again 'reviewed the literature' (including of course the *Proceedings* of past WEDC Conferences) and found little objective research other than the Kalbermatten studies. However, notes I have made during nearly forty years of association with low-cost sanitation have been useful, as was also a lot of anecdotal information from WEDC students.

My next step was to start a postal survey, first directed at former WEDC students who have returned to their native lands. Early questionnaires covered a wide range of developing countries, but some of the less-poor countries had little interest in *low cost* systems. So I limited my list to the following countries with GNP less than $1000 per person:

Bangladesh	Kenya	Philippines
Bhutan	Lesotho	Sierra Leone
Cameroon	Malawi	Sri Lanka
Egypt	Mozambique	Sudan
Ethiopia	Myanmar	Swaziland
The Gambia	Nepal	Tanzania
Ghana	Nigeria	Uganda
India	Pakistan	Zambia
Indonesia	PNG	Zimbabwe

I added Botswana, where there has been some interesting low cost work even though the GNP is $2040 per person. Of these 28 countries I have been to 23 at some time or another and have some familiarity with their sanitation problems.

For most poor people low cost sanitation means some form of pit latrine. In Africa and other places where most people use solid anal cleaning material such as leaves or newspaper a simple pit is common. There are many improvements and variations including the VIP, the KVIP, the SanPlat with tight-fitting lid, raised latrines, step latrines, double pits, the ROEC and borehole latrines. Where people use water for cleaning themselves (including inhabitants of the Indian sub-continent and followers of the prophet Mohammed elsewhere) the pit is usually separated from the latrine by a water seal, with various forms of pour-flush arrangement.

In general pit latrines are accepted for rural areas. Most attention on low cost sanitation has concentrated on devising technology that is suitable for scattered communities. Even more effort has gone in recent years to the software side of rural sanitation such as health education for villagers and involving village women in sanitation management.

When it comes to towns and cities there is no consensus as to the desirability of low-cost sanitation. There are wide differences of opinion. Some condemn pits, or indeed any form of on-site sanitation, as unsuitable for all urban areas. Even so-called 'appropriate technology' seems prejudiced against built-up areas. For example it is recommended that VIP latrines should be in windy places or the sun should shine on their vent pipes. Moreover there is a lot of fuss about the build up of groundwater pollution when houses are close together.

Consequently my research tends to concentrate on town-dwellers, trying to discover the extent to which objections to on-site sanitation are justified. Conversely, what forms of low cost sanitation are as satisfactory for urban as for rural populations in both the short term and long term?

A limited postal survey was carried out during 1992. Questionnaires were sent to about three hundred contacts on the WEDC mailing list, most of whom are working in some aspect of water and sanitation, although few have a major involvement with low cost sanitation. Most replies came from Africa, so deductions are not globally significant. We hope to extend the survey considerably, to get many more replies, to get information from other regions and to involve more people who are actually working in or have a special interest in low cost sanitation.

Questions which I asked included those dealing with the following aspects:

- **Types of household sanitation in use**: because of the African slant the most common type is the simple pit latrine, with WCs discharging to septic tanks coming a poor second. VIPs were reported for 11 per cent of households and other systems each had a mere handful of users.
- **Reasons for households having no latrines**: cost came top of the list; some acknowledged the problem of small plots; a few suggested that people prefer to defecate in the open; ignorance of how to construct a latrine is often a reason for doing nothing.
- **Emptying full pits** was reported for a substantial minority of households, with about equal numbers emptied by the householders themselves and by local government. Contractors do the job for about a sixth of the latrines that are emptied. Almost everywhere disposal of the contents is a problem.
- **Discharge of septic tank effluent to open drains** was only reported for a few towns.
- **Use of household latrines by children** was included because I have so often seen children old enough to look after themselves defecating indiscriminately even when their own family has a latrine. Two thirds of respondents replied that some adults object to children using adult latrines.
- **Householders' payment for latrines**: in only 29 per cent of places do the beneficiaries pay all the cost and in only 6 per cent is the cost of 'normal' latrines affordable by everyone.

The questionnaire asked about *multi-compartment latrines* – public latrines in markets and similar places; communal latrines used by people when at home where they have no household latrine; and institutional latrines in schools and similar places. Replies show that three-quarters of such latrines are cleaned by employees; a quarter by the local community themselves. In about half the towns someone is in attendance whenever the latrines are open, and in about half (not the same half) most latrines are open 24 hours a day. A fee is charged for use of some or all public latrines in about half the towns – again a different half – and in many towns this fee covers the cost of attendants' wages. In less than 30 per cent of towns is there lighting in most public latrines at night.

Multi-compartment latrines are renowned for poor maintenance, so it is not surprising that less than ten per cent of towns reported good conditions. Elsewhere latrines are dirty, floors are usually fouled with faeces and there is damage such as broken slabs and missing doors.

I am sure that much more information can be obtained by questionnaires, although without field work the replies have to rely on subjective opinions of respondents. Unreliability was demonstrated when I went to one of the cities in Africa from which the returned form had reported that five per cent of latrines were VIPs. I asked to see some of them. Five per cent meant there should be several thousand VIPs but the Council staff had difficulty in locating half a dozen and we found that most of these were not used.

Previous WEDC research on infrastructure for low income housing had been largely based on experience in the Indian sub-continent (Cotton & Franceys, 1991). From this we realized that a major obstacle to provision of latrines in some urban areas is small plot size. Some designs for twin pit pour-flush latrines are suitable for plots as small as 26 square metres (Ribeiro, 1985), but regulations may prohibit on-site sanitation with high population density. This whole question of sanitation for small plots needs thorough investigation.

WEDC has listed other topics for further study. This can be done by extending the postal survey and by fieldwork to give inter-country comparisons.

1. **Reasons for lack of household latrines**: the postal survey indicates that affordability is the main problem, followed by householders' lack of knowledge and several other factors; but some hard quantified data is needed.
2. **Fly and odour control**: more information about the performance of VIP latrines in urban areas is needed; how effective are tight-fitting lids over unvented pits?
3. **Pit emptying**: what are the best systems for emptying pits (and also vaults and septic tanks) and for the hygienic and low cost disposal of sludge?
4. **Operation of twin pits**: we plan to analyse the extent of and reasons for incorrect operation of twin pits. Some householders do not 'alternate' the pits properly.
5. **User satisfaction**: what do users like and why, problems in use and maintenance, and whether there are changes in use and satisfaction after latrines have been in use for some time.
6. **Groundwater pollution** is often used as an argument against on-site sanitation. To what extent is this

justified? There is a need for technical, economic and social comparisons of removing excreta by sewerage so that extracted groundwater is unpolluted as against accepting groundwater pollution from on-site sanitation and obtaining piped water from distant unpolluted sources.

This is an ambitious list. We realize that even if we are well sponsored WEDC can only deal with selected aspects of these topics. However, networking of research may enable a great deal to be achieved. Sharing of information already exists to some extent. For example, topic 3 in the list (emptying and disposal of solids from pits and septic tanks) is already dealt with by IRCWD in Switzerland and *WASTE* in Tanzania and the Netherlands. It is the subject of papers at this Conference by Martin Strauss and by Muller, Rijnsburger and Kirango. So there is no need for WEDC to give specific attention to emptying and disposal. If we come across useful data or ideas, we will pass them on to Martin or *WASTE*.

It is sometimes stated that low cost sanitation should be 'site-specific'. In this it is poles apart from much conventional civil engineering. In particular it is quite different from piped water supplies or conventional sewerage, whose technology is more-or-less the same everywhere.

The variety of low cost sanitation makes it a particularly interesting field to work in. It is also difficult to suggest methods that apply everywhere. What is appropriate in South Asia may be completely unsuitable for West Africa. Many countries or districts have their own widely used type of latrine. Zimbabwe boasts wide coverage of VIPs, Ghana favours the KVIP and in India there are hundreds of thousands of twin pit pour flush water-seal latrines (Roy et al, 1984). Compost latrines have only been used in any numbers in Vietnam and Guatemala (Hunt, 1986). A couple of years ago I visited Myanmar and found a commonly used latrine that I have seen nowhere else. Because both water and sticks are used for anal cleaning, this latrine has a pan like that for pour-flush latrines. A straight pipe goes from the pan to a single offset pit.

Even within one district or city satisfactory sanitation for some householders may be different from what should be done in the next village or another part of the city. The best solution for one family may not be best for the people next door. There may also be changes with time, so that what was acceptable five or ten years ago may be rejected now.

So advice on sanitation has to depend on many components. A good advisor may need to have at least some skills in social and cultural matters, in soil mechanics and hydrogeology, in construction techniques, management and economic/financial aspects.

This means that low-cost sanitation research has to relate to these same varied and site-specific matters. This in turn means that what is discovered by research on one site may have limited application elsewhere. So to get a global picture (or rather a picture that applies to the whole of 'the South') we need to have a very large number of studies covering all the varied social and cultural matters, soil mechanics and hydrogeology, construction techniques, management and economic/financial aspects.

Field studies can be very expensive and take a very long time, especially if conducted by expatriate professionals. Look, if you can get hold of it, at the report on household demand for latrines in Kumasi (Whittington et al, 1982). I noted above that the 1970s World Bank research only covered a few countries. In fact it only covered a few areas in the few countries it investigated - for example, only the Ibadan core area and Bussa New Town in Nigeria (Feachem et al, 1979).

However, there are many small-scale investigations going on all the time. Some are conducted by agencies responsible for sanitation. Some by consultants. More in number are undertaken by people in universities and colleges, for example when undergraduate or postgraduate students look at local situations for course projects. All these result in reports of one kind or another, but virtually all count as 'grey' literature because they are not published. If the information gained in these studies can be shared there would be a vast collection of worldwide information.

Sharing the results of research is one objective of networks. Dan Campbell in his paper for this Conference outlines the ways in which networking operates and introduces the Global Applied Research Network for Water Supply and Sanitation – GARNET.

GARNET is an initiative of the Water Supply and Sanitation Collaborative Council and was set up to improve the flow of applied research information in the sector among institutions and individuals, between researchers and programme implementors and between industrial and developing countries.

Figure 1. Myanmar latrine

Another advantage of networking is that it extends the possibility of working with other researchers. Collaboration between institutions in industrial and developing countries is already common as several papers presented at this Conference show. More joint efforts could be encouraged by networking.

Yet another benefit can be avoidance of duplicated effort. Too often there are several researchers doing more or less the same thing. Too often research can be criticised for 'reinventing the wheel'. It is far better if investigations can build on what others have done.

Within the global network there are about thirty topics, each with a coordinator. The following list gives the broad categories covered; some, like handpumps, have more than one coordinator, each dealing with a different aspect of the topic.

GARNET topics

Eutrophication
Groundwater pollution
Guinea worm eradication
Handpumps
Housing and health
Hygiene behaviour
Health impact assessment
Infrastructure for housing
Institutional development
Latrines
Participatory monitoring
Rainwater harvesting
Separation processes
Social science research
Solar distillation
Solar disinfection
Solid waste recycling
Toxic substances
Waste management
Wastewater reuse
Wastewater treatment
Water (unaccounted for & efficient use)
Water treatment

Those actively involved in applied research are urged to join GARNET. I will be pleased to answer enquiries and to put enquirers in touch with the appropriate topic network coordinator.

References

Cotton, Andrew and Franceys, Richard, *Services for shelter*, Liverpool University Press, Liverpool, 1991.

Feachem, R.G., Mara, D.D. and Iwugo, K.O. (Feachem et al), 'Alternative sanitation technologies for urban areas in Africa', *Public Utilities Report No. RES 22*, The World Bank, Washington DC, 1979.

Hunt, Steven, 'Lucrative latrines', *IDRC Reports*, 15, 4, October, pp 13, 1986.

Kalbermatten, Julius, Gunnerson and Mara, *Appropriate technology for water supply and sanitation* (12 volumes), The World Bank, Washington DC, 1981.

Ribeiro, Edgar F., *Improved sanitation and environmental health conditions: an evaluation of Sulabh International's low cost sanitation project in Bihar*, Sulabh International, Patna, 1985.

Roy, A.K., Chatterjee, P.K., Gupta, K.N., Khare, S.T., Rau, B.B., and Singh, R.S., (Roy et al), 'Manual on the design, construction and maintenance of low-cost pour-flush waterseal latrines in India'. *TAG Technical Note Number 10*, The World Bank, Washington DC, 1984.

Whittington, Dale, Lauria, Donald T., Wright, Albert M., Choe, Kyeongae, Hughes, Jeffrey A., and Swarna, Venkateswarlu, (Whittington et al), 'Household demand for improved sanitation services: a case study of Kumasi, Ghana', *Water and sanitation report 3*, UNDP-World Bank Water and Sanitation Program, The World Bank, Washington DC, 1992.

Part of the preliminary research undertaken by WEDC was funded by a research contract with ODA, whose support is gratefully acknowledged with hopeful anticipation of backing for the next phase of the investigation. ODA is also funding operation of the GARNET Global Network Centre at WEDC.

Treatment of sludges from on-site sanitation

Martin Strauss

IN MANY TOWNS and cities of developing countries, the disposal or use of sludges from on-site sanitation systems, i.e. from septic tanks, latrines, and aqua privies constitutes a huge problem. In the majority of situations, the sludges, which are collected and hauled by emptying vehicles, are dumped at shortest possible distances from the city, thereby causing serious health threats and environmental damage. With the continuing implementation of latrine programmes in urban areas, the problem becomes larger year-by-year. The lack of simple and low-cost treatment solutions has, in many cases, prevented authorities and enterprises from trying to tackle the problem. The International Reference Centre for Waste Disposal, IRCWD, has thus started an R+D project, the objective of which is to find sustainable processes and technologies for the treatment of faecal sludges. It would much welcome information on faecal sludge treatment practices and ongoing research. What is being presented in this paper is a contribution for discussion rather than a presentation of proven and tested solutions.

Factors affecting process choice

An important question is whether the effluents or products from a selected faecal sludge treatment installation are to be used in agriculture or in aquaculture, or whether their final destination is the discharge into the environment without prior use, e.g. into a receiving water body, or on land in the case of landfilling. In the case of use of the end products, hygiene standards such as faecal coliform or helminth egg concentrations will be important, while in the case of discharge, parameters such as organic loads (BOD or COD, e.g.) or nutrients (P and N) will be more relevant. Economic, sociocultural, institutional, climatic, and geological aspects have to be considered concurrently with the above criteria.

Faecal sludge characteristics

Table 1 lists characteristics of septage and latrine sludges reported in published and unpublished literature.

Data about septage quality are relatively abundant whereas only few data are reported about sludges from the various types of latrines. Septage and latrine sludges usually exhibit very poor settleability (U.S. EPA, 1984).

Treatment options

Simple and low-cost: necessary but not sufficient!

However adapted and appropriate a chosen solution might be, engineers and planners must be aware of the

Table 1. Characteristics of sludge from on-site sanitation systems

BOD_5 (mg/l)	COD (mg/l)	Total solids(%)	TKN (mg/l)	Eggs (no./l)	Country	Reference
Septage:						
3,100-5,900	16-60,000	1.1-3.9	410-820		U.S.	EPA (1980)
7,000[1]	15,000	4	700		U.S. (design)	EPA (1984)
		2-4			Asia	Pescod (1971)
680	8,100				Ghana	Accra Waste Man't. Dept. (1992)
1,600	5,750				Jordan	Al Salem (1985)
	24,400	4.7	544 (N_{tot})			Jak. Sewer.+ San. Project (1982)
2,500- 3,000		1.5-2.5			Thailand (BGK)	Edwards et al. (1987)
3-6,000	17-23,000	2-2,5	6-6,500		S. Korea	Yao (1978)
3-5,000	8-15,000	2-3	5-6,000	40-100	Japan	Yao (1978)
Latrine sludges:						
15-18,000	26-33,000	1.2 - 3	5-6,000	18-360,000	China	Shiru + Bo (1990)
30,000	50,000	1.2	450(N_{tot})	54,800	China (Jangxi)	Shiru + Bo (1990)
			2,800 - 4,750		Shanghai	Edwards (1992)
		15-54			Tanzania	Hawkins (1981)
				1,000 (stored)	Guatemala	CEMAT (1992)

fact that unless a suitable institutional framework exists or is set up through which the installations will responsibly be taken care of, and unless those in charge of operation, maintenance and management of such installations are properly instructed, trained and their salaries regularly paid, even the simplest treatment system will fail! What is needed is the awareness and recognition that anybody dealing with shit carries at least as much prestige as a person dealing with potable water supply, road construction or being a doctor, lawyer or minister!

Options overview

One basic distinction which can be made in classifying the treatment options is between separate treatment of faecal sludges, i.e. without mixing them with wastewater, and co-treatment, which consists in treating septage or latrine sludges jointly with municipal wastewater or with solid wastes. Another distinction is between processes which lead to a degradation of the organic matter, and processes which lead to a direct dewatering or drying without much biochemical decomposition. Thus, the options list presents itself as follows:

Separate treatment

Direct dewatering or drying
1 Drying (evaporation) lagoons
2 Drying beds (providing evaporation and drainage)

Schemes providing degradation of the organic matter
3 Solids separation (settling or thickening) (with subsequent dewatering/drying of the separated solids and treatment of the supernatant liquid)
4 Stabilization ponds (solids separation + partial or complete liquid treatment) + dewatering/drying of the separated solids
5 Anaerobic digestion + dewatering/drying

Co-treatment
6 With wastewater
7 With sludge from sewage treatment plants
8 Thermophilic composting with refuse or other bulking/compostable material

Below, each option is briefly described, including a functional sketch. Further to this, important factors, criteria, unresolved questions and potential problems which might be associated with the particular option are listed or discussed. Where relevant, related literature is cited.

Separate treatment

Drying (evaporation) lagoons (Fig.1a)

Process description:
- Multiple lagoons operated in parallel
- Max. filling depth with fresh sludge: 30 cm (layers thicker than this will dry at the surface only and remain jelly-like below)
- Dewatering/drying of subsequent 30 cm-layers in the same lagoon until the useful depth is reached
- Emptying of the full pond manually or by front end loader, e.g.
- Decanting of supernatant theoretically possible (leading to a faster dewatering/drying) but practically difficult to operate

Factors, criteria, researchable questions:
- Treatment criteria: % solids of cake, hygienic quality (Ascaris eggs if ascariasis endemic in area)
- Climate (wet vs. dry periods) determines suitability of process; effect of rainfall?
- Will "old", dried layers of sludge become fully wetted and remain jell-like if subsequent layers of fresh sludge are being added?
- What level of sludge dryness is optimum to allow easy removal from the pond?
- Odor development and prevention
- Degradation of organics
- Treatment/disposal of supernatant if decantation is practiced?
- Method of removal of dried sludge from the pond

References:
Pescod, 1971

In Maseru, Lesotho, drying/evaporation lagoons for the treatment of pit latrine sludge will become operational in autumn 1993. A simple monitoring programme will allow to determine the main operational parameters (% total solids, organic/mineral content of the solids, Ascaris eggs) and adapt the mode of operation if necessary. One of the ponds which were not originally conceived for drying operation will be partitioned to allow filling to only 30 cm. Operation with successive 30-cm-layers of drying sludge will also be tried.

Drying Beds (Fig.1b)

Process description:
- Open, shallow containments with constructed underdrains, allowing both evaporation and seepage/drainage (in contrast to drying lagoons which allow for evaporation only)
- Batch operation in single layers of 25-30 cm of sludge

Factors, criteria, researchable questions:
- Climate: dry vs. wet periods; effect of rainfall
- Drying rate faster than in drying lagoons? - Odors
- Drying period to allow for worm egg die-off and easy removal and handling
- Treatment of drained liquid: quantities and quality of liquid, best disposal?
- Need for buffering lagoons for storing fresh sludge during wet periods?

References:
Pescod, 1971

Drying beds are or have been widely used throughout Europe and North America for dewatering sludges from sewage treatment plants. Like lagoons, drying beds require much space. In some areas, the technology therefore had to be replaced by other dewatering processes (centrifuging, filter pressing), as sludge quantities increased with the growing of the cities and the increased level of treatment and with the city-near land becoming increasingly scarce and expensive.

Settling/thickening (Fig.1c):

Process description:
- Use of sedimentation tanks or thickeners for the separation of the settleable and floatable solids
- Batch operation in multiple units since continuous sludge and scum removal requires relatively sophisticated mechanical equipment

Factors, criteria, researchable questions:
- Optimum (surface) liquid and solids loading rates
- Sludge and scum accumulation rates and optimum operational cycles
- Supernatant characteristics and treatment
- Sludge and scum characteristics and treatment
- Mode of tank emptying

Parallel units of settling tanks have been used in two septage treatment plants in Accra, Ghana, over the last few years. The tanks are batch-operated. One of the short sides of each unit is constructed as a ramp. This allows the access by front-end loaders for emptying. In the timber-rich zone of southern Ghana, sawdust is a plentiful waste product from the timber industry. It is available free of cost and used as a bulking agent and carbon source for the composting of the sludge removed from the settling tanks. The supernatant liquid is further treated in a series of waste stabilisation ponds. Only few operational and performance data have been collected to date. An in-depth process monitoring and evaluation is being planned to take place in 1993-94.

Stabilisation ponds (Fig. 1d)

Process description:
- Liquid-solids separation and liquid "stabilisation" (biochemical decomposition and pathogen die-off) in a series of ponds
- Removal and treatment of the sludge produced in the first and second pond
- Alternative: ponds preceeded by settling tanks to cater for bulk of solids separation

Factors, criteria, researchable questions:
- Effluent "standard" for discharge or for reuse?
- Optimum pond loading rates (high-strength liquid !)
- Dilution water required to make up for evaporation losses?
- Retention time required for effective organics removal vs. time required for pathogen removal
- Sludge accumulation rates and treatability

References:
McGarry and Pescod, 1970; Mara and Pearson, 1986

Waste stabilisation ponds are rather widely used for the treatment of septage and other faecal sludges. However, this treatment usually consists in the joint treatment with wastewater (see the section on co-treatment below). To date, only few pond systems have been devised which treat exclusively faecal sludges or the liquid portion thereof. The author is aware of a few systems in Ghana (3 in operation, 1 being planned) and one recently constructed for the city of Cotonou, Benin. Even after removal of the settleable solids, concentrations of the organic matter in the liquid fraction of faecal sludges are 10-20 times higher than in normal wastewater. Pond schemes treating such sludges or their liquid fraction would thus consist of a series of several anaerobic ponds before concentrations low enough for facultative pond conditions can be attained. McGarry and Pescod (1970) recommend to use the highest possible loading on successive anaerobic ponds so as to minimize the total pond surface area.

Anaerobic digestion (Fig. 1e):

Process description:
- Digestion of the sludge in an anaerobic digester prior to dewatering/drying
- Potential for methane gas recovery
- Solids-liquid separation in a second digester or in a sludge lagoon
- Treatment of the supernatant e.g. in a waste stabilisation pond, in an anaerobic filter or in a UASB (upflow anaerobic sludge blanket) reactor

Factors, criteria, researchable questions:
- Technology relatively capital-intensive and requiring skilled operating personnel; applicable in economically and technologically fairly advanced countries, only

References:
Snell, 1943; Pescod, 1979

Anaerobic digestion is a process widely used e.g. in Japan and in South Korea for the treatment of septage and other faecal sludges in so-called nightsoil treatment plants. In South Korea, the sludges are dewatered after digestion and either landfilled or co-composted with organic waste from farms in small rural-based composting plants. The supernatant from the anaerobic digestion is diluted with fresh water and treated by activated sludge.

Co-treatment

With Wastewater (Fig. 2a):

Process description:
- Co-treatment in waste stabilisation pond schemes or in so-called conventional sewage treatment plants (activated sludge or trickling filter)
- Either: plant/scheme was originally designed to take on a given mix of wastewater and faecal sludge; or: practice of blending sludges with wastewater has just established itself over the years

Factors, criteria, researchable questions:
- In the case of adding faecal sludge to a wastewater treatment system which was not originally designed for co-treatment:
  How are the operation and the effluent quality of the plant affected by the sludge addition? What are maximum added organic loads before a plant "fails"?

How much more sludge is being produced and how is its treatability affected?
- In the case of designed co-treatment:
Is it more economic and technically feasible to co-treat faecal sludges or to treat them separately? What are relevant criteria which determine the maximum percentage of faecal sludge in the plant inflow?

References:
Jewell, Howley and Perrin, 1975; U.S. Environmental Protection Agency, 1984

In many situations where wastewater treatment schemes exist and are functioning, septage or latrine sludge are added to either the last manhole upstream of the treatment works, at other points in the sewerage system, at specifically designed receiving or storage installations at the headworks of an STP, or directly into ponds. In the United States, the practice of co-treatment in activated sludge or trickling filter plants is fairly common. Problems are reported because STPs get overloaded and much septage disposal still goes on uncontrolled (U.S. Environmental Protection Agency, 1984; Mancl, 1986). Examples of co-treatment in ponds are the city of Gaborone as well as other towns in Botswana, Dar-es-Salaam and various towns in Malawi. The Al Samra waste stabilisation pond scheme treating the sewage from the city of Amman, Jordan, receives in the order of 60,000 m3/day of wastewater-cum-septage, 10 % of which is septage collected in the Greater Amman Area.

With STP Sludge (Fig. 2b):

Process description:
- Where anaerobic digestion is used already for the treatment of sewage treatment plant (STP) sludge, such installations might be suited to co-treat faecal sludges
- STP sludge and faecal sludge flows are blended and digested in a single or two-stage digesting system
- The supernatant is co-treated with the wastewater
- Drying or co-composting of the digested sludge

Factors, criteria, researchable questions:
- How do faecal sludges affect the process? Maximum mixing ratios? Gas production? Dewaterability of mixed digested sludge as opposed to STP or faecal sludges alone?
- In a plant not originally designed to receive this extra sludge stream: what is the maximum additional hydraulic or solids load which still permits proper operation and treatment?
- Effect on supernatant quality, quantity and treatability? Co-composting (Fig. 2c):

Process description:
- Turnable or static, forced aeration windrows made up from appropriate mixtures of fresh or settled or dewatered faecal sludges and refuse
- Attainment of thermophilic temperatures (50-60 °C) which lead to rapid inactivation of pathogenic organisms
- Mixing of faecal sludge (high in nitrogen) with refuse (high in carbon) allows for more optimum C:N ratios than if either of the wastes is composted separately.

Factors, criteria, researchable questions:
- Storage and pretreatment requirements for the faecal sludge?
- Mode of mixing the two components; health risks in handling
- Process limitations as a function of mixing ratios, humidity, C:N ratios; process control
- Appropriate degree of manual vs. mechanized operation
- Quality of the end product
- Compost marketing
- Compostability of fresh sludge (e.g. nightsoil from bucket latrines) vs. partially or largely decomposed faecal sludge (e.g. from septic tanks or from pit latrines)

References:
Scott, 1952; Shuval, Gunnerson and Julius, 1981; Obeng and Wright, 1987; La Trobe and Ross, 1992

Co-composting is both a traditional process as well as a fairly recent "discovery" being tried in a few places. In China, India, Malaya, Singapore and Nigeria, e.g., co-composting is being practiced for at least several decades already. Nightsoil is co-composted with either refuse and/or other organic or bulking material. Mixing ratios are in the order of 1:5 - 1:10 (sludge : added material) on a wet weight basis if underwatered sludge is used. With dewatered sludge and woodchips the ratio can be increased to as much as 1:1.5.

An example of a very recent operation is the installation at Rini near Grahamstown in Cape Province, South Africa. There, the refuse and bucket latrine sludge from a community of 100,000 are co-composted in a simply mechanized plant using forced-aearated, static windrows. The nightsoil is pre-settled and then hosed on to the windrow as the garbage is being heaped up. On a volume basis, the mixing ratio is approx. 1:10. The process is controlled by the temperatures developing within the piles. 55 °C are reached and the windrows are left to react for 3 weeks. After composting, the mixture is being sieved and the rejects landfilled. The compost is used by the Grahamstown garden department.

References

Accra Waste Management Department (1992). Personal communication.

Jewell, W.J., Howley, J.B., Perrin, D.R. (1975). Design Guidelines for Septic Tank Sludge Treatment and Disposal. Progress in Water Technology, 7, no.2, pp. 191-205.

Al Salem, S. S. (1985). Evaluation of Onsite Sewage Disposal Systems in Jordan. Unpublished.

CEMAT, Centro Mesoamericano de Estudios sobre Tecnologia Apropiada (1992). Control and Monitoring of the Dissemination of Dry Alkaline Fertilizer Latrines (DAFL) as an Agricultural and Sanitary Alternative in Guatemala. Unpublished.

Edwards, P. (1992). Reuse of Human Wastes in Aquaculture - A Technical Review. UNDP-World Bank Water and Sanitation Program. 350 pp., 470 ref.

Eikum, A.S. and Paulsrud, B. (1986). Treatment of Septage - Scandinavian Practice. Water Science and Technology, 18, pp. 63-70.

Hawkins, P. (1982). Pit Emptying Services. IRCWD, International Reference Centre for Waste Disposal, Duebendorf, Switzerland. Unpublished, 32 pp.

Jakarta Sewerage and Sanitation Project (Encona and Alpinconsult).(1982). Vol. 8, Sanitation Study. Unpublished. September.

Mancl, K. (1986). Septage - An Overlooked Waste Dispsosal Problem. Water Poll. Control Association of Pennsylvania Magazine, 17(6), pp. 6-8.

Mara, D.D., Pearson, H. (1986). Artificial Freshwater Environment: Waste Stabilization Ponds - Chapter IV.B, Nightsoil Ponds. In: Biotechnology (Rehm and Reed, Eds.), Vol. 8

McGarry, M.G. and Pescod, M.B. (1970). Stabilization Pond Design Criteria for Tropical Areas. Second International Lagoon Symposium, Kansas City. June. 22 p

Obeng, L.A. and Wright, F.W. (1987). The Co-composting of Domestic Solid and Human Wastes.. Integrated Resource Recovery, World Bank Technical Paper no. 57. 101 p.

Pescod, M.B. (1971). Sludge Handling and Disposal in Tropical Developing Countries. Journal Water Pollution Control Federation, 43 (4), 555-570. (identical to reference Pescod Jan. 1970).

Scott, J.C. (1952). Aerobic Composting (of nightsoil) with Farm Materials. In: Health and Agriculture in China.. Faber and Faber Ltd, London. (p.159-190).

Shiru, N. and Bo, L. (1991). Health Assessment of Nightsoil and Wastewater Reuse in Agriculture and Aquaculture: Final Report. Chinese Academy of Preventive Medicine, Institute of Environmental Health and Engineering, Beijing (Technical Services Agreement WP7ICP7RUD/001/RB/88-310, WHO Western Pacific Regional Centre for the Promotion of Environmental Planning and Applied Studies, PEPAS). 24 pp.

Shuval, H.I., Gunnerson, Ch.G., Julius, D.S. (1981). Nightsoil Composting. Appropriate Technology for Water Supply and Sanitation. World Bank. 81 pp.

U.S. Environmental Protection Agency (Roy F. Weston, Cons. eng.) (1984). Handbook - Septage Treatment and Disposal. EPA-625/6-84-009. U.S. EPA Municipal Environmental Research Laboratory, Cincinnati, Ohio 45268. October. 300 pp., 73 ref.

U.S. Environmental Protection Agency (Clements, E.V., Otis, R.J. et al.) (1980). Design Manual - Onsite Wastewater Treatment and Disposal Systems. EPA-625/1-80-012. U.S. EPA Municipal Environmental Research Laboratory, Cincinnati, Ohio 45268.

Yao, K.M. (1978). Utilization of Nightsoil Treatment End-Products. (South Korea). Unpublished.

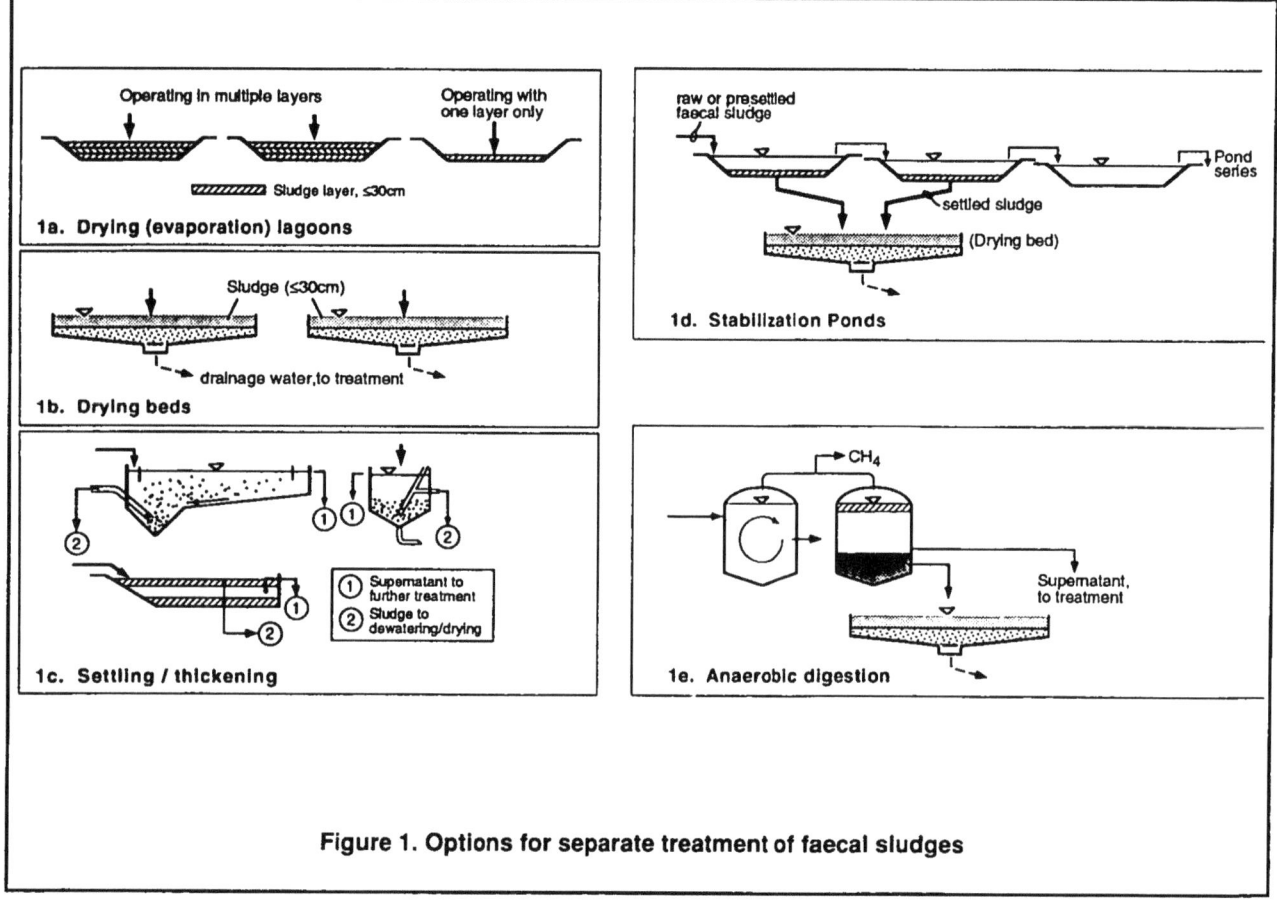

Figure 1. Options for separate treatment of faecal sludges

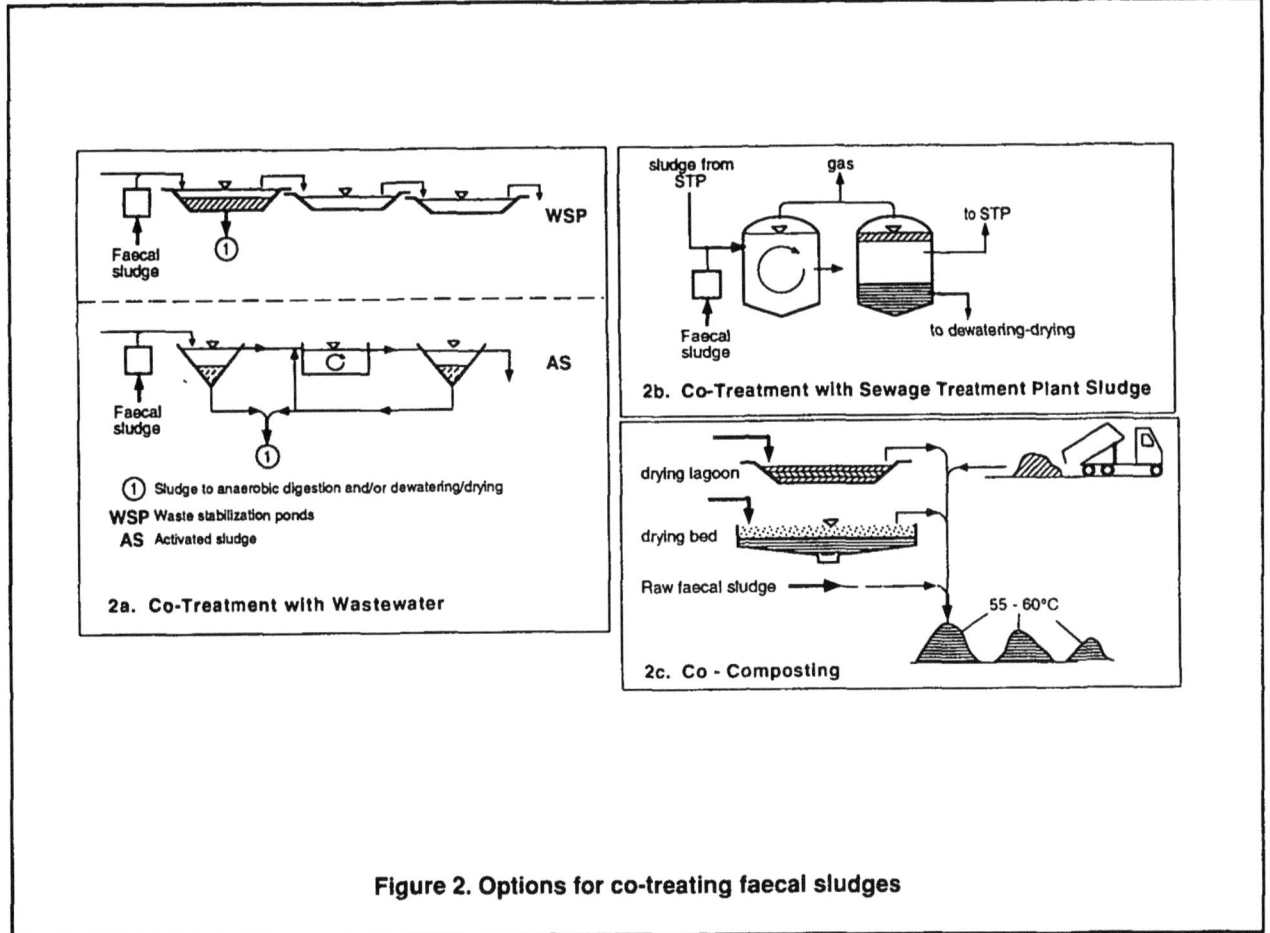

Figure 2. Options for co-treating faecal sludges

SECTION 7

SOLID WASTE MANAGEMENT

Informal sector waste recycling

Syed Mansoor Ali, Andrew Cotton and Adrian Coad

THE PROBLEMS OF solid waste management, and their solutions, are different in developing countries as compared to the developed world. The options available vary with the prevailing socio-economic and political atmosphere. In developed countries systems of collection and disposal are quite efficient and so effort is concentrated on aspects such as recycling, environmentally acceptable disposal methods, landfill gas utilisation and source separation. On the other hand, in most of the large urban centres of developing countries there are inadequacies even in the collection and transportation of waste, there are very few environmentally controlled disposal sites, and often the official efforts to develop recycling have been futile. The causes of these shortcomings can often be traced to the nature of the waste, financial constraints, problems of inadequate infrastructure, high rates of population growth, and the lack of public and political awareness. It is clear that there is a need for tailor-made and appropriate solutions to the problems of solid waste management in developing countries.

It is important to study current practices which have evolved independently of the official refuse management practices and which are helpful to the existing official systems. One such system operating in most of the large cities of developing countries is informal sector solid waste recycling (ISSWR). Such systems usually exist in parallel to the private formal and official sectors; however the objectives of both the sectors are different. These informal systems are usually unorganised, unaccepted by, and unknown to the official sectors. There are a few situations where informal systems have been studied and accepted by the government sector; one such example is the acceptance of informal systems of housing (squatter settlements) when government has failed to fulfil the housing needs of the poor in developing countries. At present, the informal sector is playing a vital role in addressing the economic needs of urban poor.

ISSWR is supplementing the service provided by the municipal sector and reducing the quantities of waste that are a burden on the environment. This paper is based on data collected in Karachi, Pakistan as a part of a postgraduate research project. It discusses various options for improvement which are also applicable to other developing countries. It also reviews the existing role of the informal sector in solid waste recycling which provides major assistance to the government sector in terms of reductions in quantities, employment and provision of cheap raw materials.

There are also hazards attached with this activity which are discussed. An in-depth understanding of the present will be of great help in formulating possible options for improving the future.

The existing system

Karachi is a typical case of a large city in a developing country. It is the largest city of Pakistan with an area of 1800 sq km and population of 8 to 9 million (KDA, 1989). Rapid rural to urban migration, coupled with increased activity of commerce and trade, has created tremendous civic problems including inadequate solid waste management. The existing system is managed by Karachi Municipal Corporation (KMC) but in spite of great efforts by KMC the system is inadequate and inefficient. The total daily generated domestic waste is 5000 tons and KMC is unable to collect even 50% of this waste (NESPAK, 1992). There are no planned landfill sites and open burning is quite common. The expenditures are increasing at an alarming rate; the yearly expenditure in 1990-91 was Rs 275 million (£7.0 million) and this has increased to Rs 400 million (£10.0 million) in the year 1992-93 (KMC 1991 to 1993). Orthodox engineering solutions call for improved vehicle designs, better containers, house to house collection etc., but all these efforts would need huge expenditures whereas existing income from taxes is very low. Under the present circumstances it is important to study economically possible solutions and improve existing self supporting practices beneficial to waste management and the environment.

One such system is the chain of private informal and formal elements involved in separation, sale, purchase, reprocessing, reuse and recycling of solid waste. (NTCS 1992).

The informal sector

Under the present complex socio-economic environment in developing countries it is difficult to isolate and define any informal sectors. Similarly Informal Sector Solid Waste Recycling (ISSWR) comprises different unofficial, private and informal elements. The complete chain of this system is shown in Figure 1. The domestic waste generated is first reduced by housewives, who separate the resaleable components such as bottles, cans, plastics, bread, etc for resale. These components are stored until sold to the itinerant street collectors. The separated quantities vary with the income groups in the range of 1 to 2 kg/family/day. This practice is common among 85% of the families in the city. The figure for the weight recycled also includes heavier materials such as glass and iron scrap.

The itinerant street collectors move from one street to another pushing their four wheeled carts and purchasing materials from houses. There are 16000 street collectors in Karachi each purchasing about 20 - 30 kg of separated waste daily. In this way a total of 500 tons of waste is separated by housewives daily. The most common material sold to street collectors is ferrous metal, followed by bread in low income areas and paper in high income areas. These components are then sold by street collectors to the middle dealers, whose shops are easily accessible in residential areas. The middle dealers are those who purchase all the materials from street collectors and also provide them with their push carts, and other services such as loans, protection from Police, etc. There are 800 middle dealers scattered throughout Karachi. After some processing, the waste goes through a stream of main dealers, finally reaching formal and informal recycling industries.

After the housewives have separated the recyclable material, the waste enters the municipal waste management stream, but it continues to be separated at various levels, reducing the quantities at each stage.

In the municipal system the waste is first conveyed by sweepers from the house to the collection points, with some sorting taking place during this process. They have limited time to sort through the refuse, and usually have offers to clean houses which are better paid. Thus they separate about 200 tons of resaleable components daily. The waste is collected by sweepers at collection points for municipal vehicles to come. During that period the transfer points are invaded by scavengers for whom sorting is a full time occupation. There are about 20,000 scavengers in Karachi and they are able to separate about 1000 tons of resaleable components daily. The type of materials separated by scavengers are different from those separated at the household level; it is more than 80% paper. The quality is also poor because it has been mixed with wet materials and consequently its value is low. Scavengers sell their collected waste to their group leaders, and it ultimately reaches the main dealers and recycling industries.

There are various end uses of such separated wastes. For example paper is converted into paper board, aluminium cans are smelted to pure aluminium metal, plastic is converted into shoes, slippers etc and ferrous metals are melted and re-rolled at large foundries.

ISSWR makes significant contributions to the economy of Karachi:

i. Quantities of waste are reduced by at least 10% in collection and 30% in transportation.
ii. Provision of full-time employment to at least 40,000 people.
iii. Supplementing the income of at least 1 million families.
iv. Supply of cheap raw materials to large and small industries.

There are also hazards attributed to this activity. For example, hazards are associated with the trade in empty chemical drums, the recovery of lead metal from used automobile batteries and its smelting, the separation of infected items from hospital waste, and the poor working conditions in small recycling industries.

Very little effort has so far been made to study and understand ISSWR. Municipal officers and the public have little awareness about it. They receive complaints about scavengers since they scatter and incinerate waste during collection. It appears that they do not believe that street collectors and scavengers really do contribute significantly to reducing the quantity of solid waste. (RAHAT ALI KHAN, 1993). The municipal officers support the concept of recycling, but their plans are at the level of installing compost plants, generating energy from waste, manufacturing fuels from waste etc.

ISSWR is currently very well established, but any over-ambitious plan by municipal authorities towards formal municipal recycling, such as the introduction of closed containers, the establishment of transfer-cum-sorting stations etc, may disturb the delicate balance of the existing set-up and imperil the vast efforts at recycling which are being undertaken by the informal sector.

It is now very important that municipal authorities in developing countries should seriously consider how to increase the benefit derived from the extensive activities of ISSWR. The sector may be upgraded and supported by government agencies for improved solid waste management.

Conclusions and recommendations

On the basis of the above discussion the following general conclusions may be drawn:

1. There is an immediate need to study the extent and potential of ISSWR in developing countries and observe whether it can be improved and upgraded to increase recycling and reduce waste quantities.
2. The hazards due to ISSWR are quite similar to any formal activity. There is a need to broaden the scope of existing legislation to include ISSWR and their honest and effective implementation is necessary to control such hazards.
3. Municipal authorities may start pilot projects to integrate ISSWR with the formal system of waste management. Similar experiments have been undertaken in other cities of developing countries, notably in Cairo, Jakarta and Manila. It is worthwhile to study the success and failure points of such systems before planning pilot projects in other cities.
4. There is a need to launch awareness programmes for public and municipal officers about the role, activities, potential and hazards of ISSWR.
5. Careful and stepwise interventions for ISSWR are envisaged at different levels. These plans should not change the informal nature of the system and the implementing agency should act merely as a catalyst for some activities
6. Under long term plans, tax subsidies and other benefits may be extended to formal and informal sector waste recycling industries as their role is crucial in creating a market demand for recycled materials.

References

KDA (1989), "Infrastructure Sector Profiles". Karachi Development Authority, Master Plan and Environmental Control Department.

KMC (1991 to 1993), "Municipal Committees Budgets". Karachi Municipal Corporation, Karachi.

NESPAK (1992), "Detailed Design and Preparation of Tender Documents for Solid Waste Management". Interim Report. Karachi Metropolitan Corporation, Karachi.

NTCS (1992), "Promotion of Waste Reuse and Recycling in Developing Countries". Final Report for United Nations Centre for Human Settlements, (Habitat), Nairobi.

RAHAT ALI KHAN (1993), "Personal Communications". Chief Engineer, Karachi Metropolitan Corporation, Karachi.

SANYAL, B. (1988), "The Urban Informal Sector Revisited", Third World Planning Review, Vol 10 No 1, Liverpool University Press.

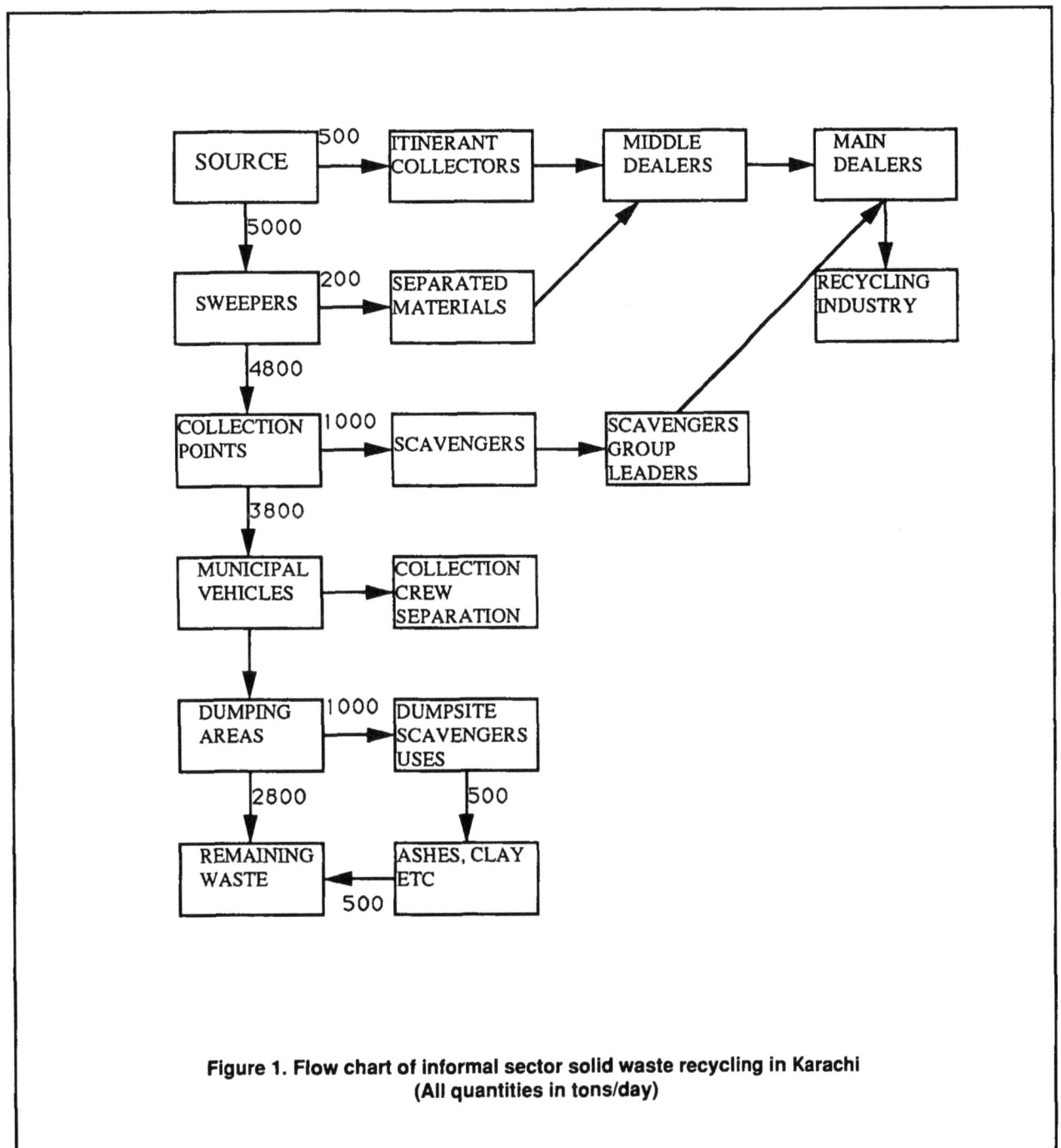

Figure 1. Flow chart of informal sector solid waste recycling in Karachi
(All quantities in tons/day)

Small enterprises for solid waste recycling

Inge Lardinois and Arnold van de Klundert

THIS PAPER IS based on the results of a comprehensive research into the great variety of products made, markets covered and technologies used in the recycling of urban solid waste by small-scale enterprises in six selected cities in the South. The research meant to seek and develop options for these small recycling enterprises to create new and more employment, but also to propose alternative solutions to the increasing problem of the removal of the urban solid waste.

WASTE Consultants has coordinated this research project, whereas the research itself has been carried out by local consultants[1] in the following six cities: Manila, Calcutta, Cairo, Nairobi, Bamako and Accra. This paper is based on their findings[2].

By describing the recycling of three materials namely, plastic, rubber and batteries, some insight will be provided into factors which are of vital importance for the economic feasibility and the scale of the recycling activities. It also shows in which way the product quality and thus marketability could be improved. Finally the need to reduce environmental hazards is stressed.

Plastic recycling

Plastic is a relatively modern material, widely used in the North and increasingly so in the South. In the North the recycling of plastic is rarely done: identification and separation of the various plastics is a technically complex and expensive process. The situation is different in many Southern countries. An ample supply of cheap labour and an increasing demand for low-priced products have resulted in a rapid growth of a labour intensive recycling industry: in Cairo more than 400 small enterprises exist, which recycle approximately 70% of the waste plastic generated and offer a wide variety of employment.

In Egypt, plastic was one of the first non-organic waste products to be recycled. This is due primarily to four factors: the fluctuating price of imported plastic polymers, the rising rate of exchange for hard currency, local inflation and the rising value of plastic waste. The plastic industry is growing fast in Egypt. The Egyptian laws prohibit the import of most plastic products, with the exception of certain items of necessity such as spare parts. Domestic production of plastic polymers is encouraged. The market for recovered polymers is highly competitive and volatile. However, despite this market volatility, prices have been increasing fourfold since 1981.

The recycling of plastic is a process in which many people are involved in the various stages of the process from 'waste' till final 'product'. The waste plastic has to be collected from dump sites or household storages, sorted out according to colour and type, washed and shredded, granulated and finally moulded into a product such as tubes, shoe soles, combs and plugs. Every additional step in the process adds value to the product and consequently generates additional income.

In the cities researched the state of the art of the applied technology however differs considerably. In Nairobi plastic is only collected, partly sorted and then sold to large processing companies. In Cairo, Manila and Calcutta all stages of the process are present. Small entrepreneurs process the waste plastic and produce low-priced products. The technology applied is adapted to local circumstances and capabilities. Local machine shops are able to construct and maintain rather advanced extruders, driven by lorry gearboxes and car tyre v-belts.

In sorting and washing the plastic there was another striking difference. Sticky liquids, solid matters mixed with the plastic or still inside containers make the sorting of plastic a dirty job. In India, however, this problem was tackled: in concrete basins the waste plastic is washed, while the water is pumped around by a small electrical engine. The washed plastic is first dried in a rotating drum made of welded mesh and then spread on a field to be fully dried by the sun. At this stadium especially women come in to sort the plastic. This different approach, first washing and then sorting, has improved working conditions considerably.

Rubber recycling

Waste rubber is one of the most important materials to be recycled in Bamako in Mali. The level of the recycling is however very basic. Relatively simple hand tools are used to cut and punch the rubber tyres and tubes into springs, pulling ropes, sandals and v-belts. Products which are a common sight in most street markets in Southern cities.

In the other cities such as Manila and Calcutta the rubber processing is more advanced and often mechanized. The waste rubber is first devulcanized, a process to remove elasticity and make the rubber mouldable. Rubber reclaimed in this way is sold to other enterprises in the tyre industry and has a wider application than crude waste rubber. Most of the recycled rubber is used for the manufacture of moulded rubber products, like solid tyres for wheel barrows, tri-cycles, pedals for bicycles, bumper pads, vibration absorbers and so on.

There is a high degree of integration between the formal and informal rubber industry. In fact often people who were working in the formal industries start on their own; the necessary equipment (mixing mills, presses, extruders and so on) is locally produced often from inferior materials from scrap yards and discarded drive units which are available at extremely low prices, however not always conforming to material specification and safety standards. But this lowers capital investment with 75% and is consequently a major reason for the high profitability of this sector. The high rate of wear and tear or breakages are easily absorbed in day to day economics of the recycling processes.

In Accra, Ghana, on the other hand, rubber recycling is quite limited. Most of the existing retreading industries were abandoned because of the lack of capital for the rehabilitation of equipment and because of the import of used rubber tyres from Europe. There are only a few small workshops producing engine seats, bearing housings and bushings with relatively simple methods like cutting and mechanical pressing using a vice. These recycled spare parts are 8 to 10 times cheaper than the imported ones. Customers are satisfied with lower prices and faster replacement service despite lower quality as compared with original parts. The craftsman is also satisfied with a lower charge and in some cases a higher replacement turnover.

Household battery recycling

Batteries are also a more modern product which is in itself a clever source of energy, but after use poses a threat to the environment and the health of people. The use of batteries is however steadily growing in Northern but also in Southern countries. In the North a feasible recycling process hardly exists. Batteries are piled up in controlled places until a solution is found. In the South due to its specific economic conditions a recycling 'industry' has evolved.

In Egypt, the batteries are stripped down with the aid of simple tools. Valuable compounds like zinc are removed and sold to zinc foundries, where the zinc is melted and reused. The carbon pins are used as heating elements and the residue is simply thrown away. Consequently the heavy metals from the partly decomposed batteries contaminate the organic waste, eaten by animals roaming around for food.

In India, there are a number of small battery manufacturers, who make new ones out of the old elements. The used up chemicals are piled up together with fresh ingredients in old zinc containers. New paper wrappings and plastic are used to give the product a good appearance. Such local batteries are sold at less than half the price of a standard quality battery. Most of these batteries are marketed in villages and are used for operating small transistors, radios and so on. These dry cells have poor operating characteristics and poor shelf life.

Only a small number of people is employed in this recycling sector. The labour intensive method of stripping the batteries is not always feasible. For example in Manila, batteries recycling is absent. The main reason for this is the low price paid for zinc. One kilogram of zinc costs U$ 1.53. Assuming that zinc represents 5% of the battery weight, to recover one kilogram of zinc will require the recovery and processing of around 20 kilograms of household batteries. Considering that the local minimum wage is U$ 4.37 per day, to cover just the labour cost would entail recovery and processing of around 57 kilograms of batteries!

People do not consider batteries as a health threat. That is why they can be seen abandoned everywhere after use: in compost heaps, in mud walls of houses, laying in the field between the crops, children playing with it. In the Philippines batteries are even used as a disinfectant for wounds and a smell repellent in pit latrines.

Also employers are not aware of the health hazards of this enterprise. Taking the minimal income and dangerous working conditions into account, recycling of batteries is not an activity which should be encouraged. Other solutions like separated storage or alternative energy sources have to be found.

Conclusions and recommendations

Product innovation

At international level recycling products, technology and markets differ remarkably and as such offer a broad range of products. At local level recycling often results in a great supply of identical, low-quality final and semifinal products, produced under severe working conditions and serving low-income markets only.

Appropriate innovation may diversify the products and improve their quality, to broaden the market into the middle-income market, thus generating more income and employment. Improved processed waste material will fetch a higher price in the formal industries. Ultimately these improvements may even reduce importation of some products.

External influencing factors

It appears that the extent of recycling differs very much per country and depends among others on world market prices of raw materials, import regulations and government subsidies. They influence the feasibility of recycling activities.

Especially in Asian countries, the informal recycling sector thrives on formal industries. Small-scale entrepreneurs often receive their educational training there. The use of second-hand machines and the local manufacture of equipment makes huge savings on capital investments possible. Also, these formal industries form a market for semi-finalized products from the informal sector.

It is clear that these factors are of vital importance for the feasibility of recycling activities and they should be taken fully into account when considering South-South technology transfer. It also calls for policy changes at the national and international level.

Co-operation between public authorities and private entrepreneurs

Up till now the efforts of private small scale entrepreneurs have not been taken seriously in the sense of their contribution to the removal of urban solid waste, the creation of employment and likewise the savings on the use of virgin (imported) materials. Authorities still opt for advanced solutions such as compacting trucks and incineration.

Authorities should become aware of the impossibility to solve the urban solid waste problem on a public base only. Legislative and policy measures should be developed to enable co-operation between the public and the (small scale = employment) private sector.

Improvement of working conditions and environment

Although recycling contributes to a sustainable development, recycling in itself needs not to be an environmentally sound enterprise. The challenge for the future lies in trying to extend recycling activities on the one hand and trying to improve working conditions and lessen negative environmental effects on the other hand.

The nearer to the source the waste is separated, the less contaminated the raw material is and the more homogenous the semi or final product is. Separation at source also improves the working conditions of those (often women and children) handling the waste.

Exchange of information and experiences at workshop level

In different cities recycling has developed in different ways and on different levels. In some cities solutions that have been found on a workshop level and at a municipal level are worth to be disseminated and replicated (and adapted) elsewhere.

Opportunities should be created (network, exchange programme, visits) to enable a free flow and exchange of information. Particularly small scale entrepreneurs, representatives of enterprise support institutions and authorities involved at this level of recycling should participate.

1 These consultants are: EQI/Cairo, AUC/Cairo, Undugu Society/Nairobi, Ptr/Calcutta, CAPS/Manila, ABP/Ghana and GERAD/Bamako.
2 Results will be disseminated in four publications on urban solid waste recovery: 'Organic Waste', 'Plastic', 'Rubber' and 'Hazardous Waste' (Tool/Waste, Amsterdam, The Netherlands).

SECTION 8

WATER QUALITY

Monitoring pollution in Lagos Lagoon systems

Margaret E Ince and T I Ojo

THE LAGOS LAGOON is the largest in an extensive coastal system in the Gulf of Guinea (Map 1). It is linked in the east to the Lekki Lagoon and to the sea at its western end, near Lagos. Although the water is saline close to this single outlet to the sea, the Lagoon is essentially a large expanse of shallow freshwater, being fed by a number of major rivers. It is approximately 20 km wide, 50 km long and 0.5-5 m deep (mainly <2 m in the main body of the Lagoon), excepting where it is dredged to about 25 m in and around Lagos harbour (Webb, J E, 1958). The Lagos conurbation, population ~10 million, is the industrial capital of Nigeria and accounts for ~70% of the nation's industries.

Domestic and industrial wastes, from the metropolitan area, and persistent pollutants carried by rivers enter the Lagoon and affect its environmental and economic status. Research [Ajao, E A (1990); Ajayi, T O et al.(1989) and Okaye, B C O et al. (1991)] has indicated that accumulation of pollutants, such as heavy metals, are affecting the ecological balance and threatening the livelihood of local fishing communities. In addition, increased national environmental awareness has placed a high priority on maintaining the aesthetics of the Lagoon.

This paper describes briefly the sources and potential impacts of pollutants on the Lagoon, based on personal knowledge of the authors, site visits, available documentation and discussions with appropriate experts.. A pollution monitoring programme and possible actions to mitigate against pollution of the Lagoon are proposed. Funding for this environmental monitoring and protection of the Lagoon, a component of the World Bank funded (IDA) Nigeria Environmental Management Project is being provided by the UK Overseas Development Administration (ODA).

Impacts of pollution on the value of the Lagoon

The Lagoon is of great value both locally and nationally. Of national importance is the diversity of flora and fauna that it supports. It provides transport of goods through the port as well as between villages along its banks. Locally it also provides water for drinking and for agriculture; a means for disposal of wastes; a source of food, especially fish and shellfish; sand which is dredged for use in the construction industry; a storage place for timber waiting to be processed; a site for recreation and a general amenity.

The value of the Lagoon is, however, reduced by some of the activities in and around it. The absence of a sewerage system in Lagos has led to a common practice of discharging untreated domestic and industrial wastes, directly or indirectly, into the Lagoon. Wastes interfere with the self-purification processes in the Lagoon; non-biodegradable chemicals accumulate up the food chain, affecting aquatic organisms and exposing humans to increased health risks. For instance, heavy metal contamination is a health hazard both for those who dredge sand and for those eating Lagoon fish in which metals have accumulated. Another reported but not quantified impact is on fish stocks. Informal interviews with fish traders at the western end of the Lagoon indicated a reduction in both numbers and income.

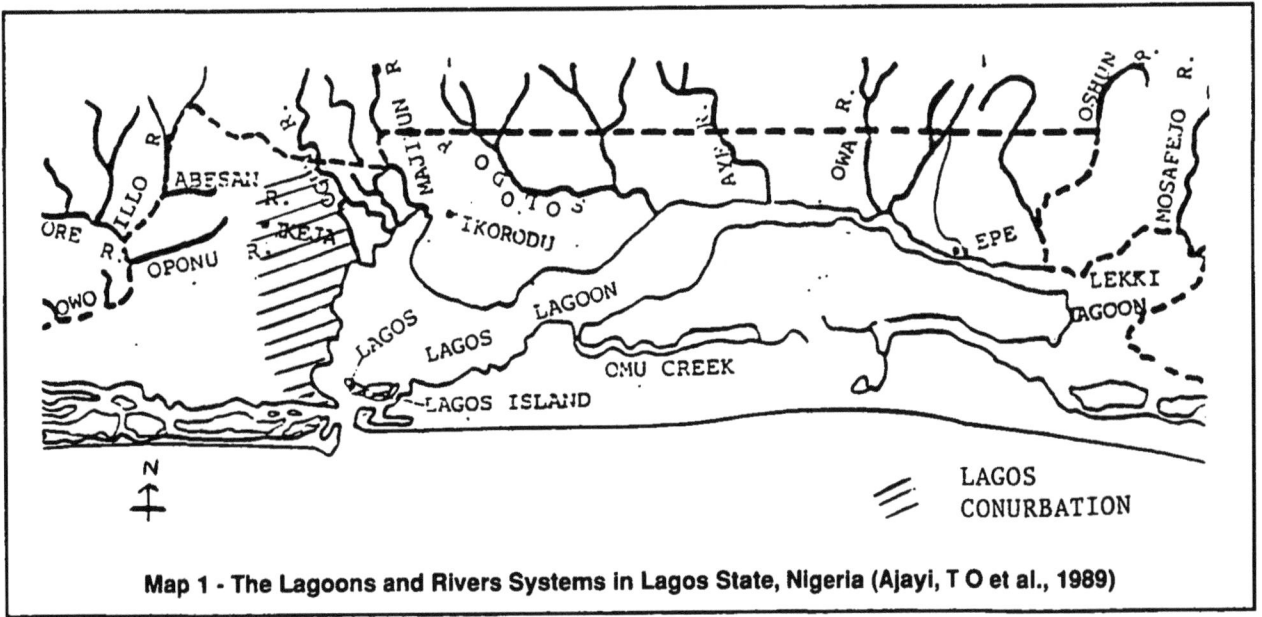

Map 1 - The Lagoons and Rivers Systems in Lagos State, Nigeria (Ajayi, T O et al., 1989)

Sources of pollution

Most (potential) sources of pollution are concentrated at the western end of the Lagoon where the greatest concentration of heavy metals has been detected (Okoye, C O et al., 1991). There is the normal wide range of urban industries including breweries; manufacture of textiles, detergents, and pharmaceuticals; abattoirs and food processing; and petroleum industries. However other industries of local importance are the small and commercial fisheries, wood processing, and shipping. The absence of sewers and controlled sites for waste disposal leads to the discharge of tanker loads of septic tank sludge directly into the Lagoon (at Iddo Jetty). Apart from its oxygen demand, this organic waste results in a concentration of faecal organisms and hence a greater public health risk for people in this area or eating (shell)fish (especially undercooked) caught here. During one site visit (February 1992) faecal coliform count greater than 30,000/ml was obtained in an inlet close to the jetty. Additional pollution may be introduced by run-off from the extensive inter-island/mainland road network.

Pollution monitoring programme

Objectives

- assessment of the present quality of the Lagoon in relation to flora, fauna and water quality;
- establishment of the hydrology of the Lagoon ;
- quantification of the health, financial and economic implications of pollution;
- devising appropriate solutions to clean and abate pollution; and
- establishing pollution control strategies to reduce pollution to acceptable levels within ten years.

Water quality monitoring

To provide baseline data ion the Lagoon, a provisional list of parameters to be monitored has been prepared. Where possible, monitoring will be carried out on site. Analysis for parameters present at low levels or requiring sophisticated equipment, eg pesticides, will be laboratory based. The number and siting of sampling points will cover the whole of the Lagoon but will be concentrated at the western end of the Lagoon and possibly close to mouths of influent rivers. Depending on preliminary findings, frequency of monitoring will be established and the range of parameters may be modified. A long-term monitoring programme will be designed and implemented. To support this component, institution capacity building (human resources development and laboratory facilities) is provided.

Effluent monitoring

To assess the scale of pollution and scope for remedial action, industrial (and septic tank) effluents will be monitored for quality and quantity. It is anticipated that consents will be established, within policy constraints, by negotiated rather than dictated agreements between industries and the enforcing agencies. Liaison between these actors, especially in relation to sequential improvements in effluent quality, should provide greater benefit in the long-term.

Conclusion

Lagos Lagoon receives pollutants from a variety of sources. The quantity and impact of this pollution is largely unquantified. The project described briefly above aims to collect data on the effluents entering the Lagoon and the current status of the Lagoon itself. It is expected that the data collected during the project will lead to the design and implementation of a long-term monitoring programme for pollutants both in effluents and the Lagoon and the development of action plans to mitigate against impacts of pollution. The final outcome will be a reduction in levels of pollution entering the Lagoon and, hopefully a reduction in pollutants already present.

Acknowledgement

The authors wish to acknowledge the financial support of the ODA, and staff of FEPA, the Nigerian Institute for Oceanography and Marine Research and Lagos State Government for their assistance with data collection that contributed to the preparation of this paper.

References

Ajao, E A (1990) PhD thesis. *"The influence of domestic and industrial effluents on populations of sessile and benthic organisms in Lagos Lagoon"*. University of Ibadan, Nigeria.

Ajayi, T O, Ayinla, A O, Udolisa, R E K, Bolande, E O and Omotoyo, N O (1989). *"Diagnostic survey of small scale capture and culture fisheries in Lagos State, Nigeria"* Technical paper 56, Nigerian Institute for Oceanography and Marine Research, Nigeria.

Dobie, P; Ince, M E and Winpenny, J T (1992). *"Nigeria: Lagos Lagoon Project"*, Overseas Development Administration.

Okoye, B C O, Afolabi, O A and Ajao, E A (1991). *"Heavy metals in the Lagos Lagoon sediments"*. International Journal Environmental Studies Vol. 37 pp 35-41.

Webb, J E (1958)."Ecology of Lagos Lagoon. The life-history of *Branchiostoma nigeriense* Webb". Phil. Trans. Roy. Soc. Bull. Vol 241, 335-353

Water quality in family wells

Grace Rukure, Shungu Mtepo and Cornelius Mukandi

STANDARDS OF DRINKING WATER and public sanitation are low in many rural communities in developing countries. Contamination of drinking water with pathogenic microorganisms maybe an important source of waterborne diseases such as bacillary dysentery, cholera and typhoid. It has been estimated that as many as 80% of all diseases in the world are associated with unsafe water.

In Zimbabwe, water quality is not a problem in urban areas, but then only 40% of the population lives in the town. A large number of family wells exist in the country, an estimated 100 000. Most of these are poorly protected and yield water of poor quality.

Since 1988, the Blair Research has been investigating ways of protecting family wells in order to enhance the quality of water they yield. Hence the inception of the Upgraded Family Well, which was a simple method of protecting family wells without the use of a handpump. In the upgraded family well emphasis has been placed on lining of the well from the bottom right up to the top, with a simple apron and run-off around the well-head which should be raised at least 300m above ground level. A simple concrete slab is then placed over the well and finally a simple improved windlass, bucket and chain arrangement is then fitted supported by two cement mortared brick pillars. The central access hole in the slab is covered with a metal tin lid. Previous work on water quality has shown that significant improvements in water quality can be made by this type of protection of the well (1,2). However, there is still insufficient scientific data available on the various components of the family well and their relative importance vis a vis water quality. The government of Zimbabwe at the moment offers a subsidy of 3 bags of cement, a windlass and a lid to each family as its contribution to this rural water supply programme. However, with the Economic Structural Adjustment Programme, this subsidy will not be there forever. What options could then be there? In many countries the windlass, which is the most expensive component of the technology, is quite unknown. Attempts have therefore to be made to investigate means of improving water quality in the absence of a windlass. Would improving the headworks only, be sufficient? If the pit was not lined, would this make a drastic difference to the water quality? This study attempts to answer some of these questions.

Materials and methods

Four communal areas were selected in Mashonaland East has a large number of family shallow wells and some of these wells already upgraded. Convenience sampling was done to select a number of family wells which were either upgraded, improved or open. Various features of the improved and open wells were noted and then these subdivided into various types according to the absence or presence of various technological components of the well.

Water samples were then collected from each well for analysis and each sample analysed for faecal contamination, pH and turbidity. The membrane filtration method using lauryl sulphate broth was employed for enumeration of *E. coli* (3).

Results

Level of contamination

This was indicated by the numbers of *E. coli* in the samples, for the different types of wells as shown on Table 1. The Upgraded Well produced water with the lowest level of *E. coli* (mean count 45/100ml) and the unimproved wells, the highest level of contamination (mean count 182 and 263/ml for Type 1 and Type 2 respectively) within the various categories of "improved" wells, the well with no cover slab, no lid, and no windlass provided the lowest quality water. Type 2, 3 and 4 wells were not significantly different.

PH and turbidity

For the turbidity and pH analysis, a total of 80 wells were examined, 16 upgraded, 32 lined, therefore improved wells and 32 unlined, therefore unimproved wells. Here it was not necessary to differentiate the improved wells into the presence or absence of certain technical components. Results obtained are as shown on Table 2.

The unupgraded, but lined wells had turbidity values significantly higher than those of the upgraded wells, but significantly lower than the unlined wells. All the lined wells had values below the maximum permissible level of 25 (Turbidity Units) for rural community water supplies. Ph values for the upgraded lined and unlined wells were not significantly different.

Discussion and conclusions

The present study is one of the few studies carried out which has wide implications and application in Zimbabwe and the subcontinent in terms of rural water supply programmes. What is obvious here of course is that any shallow well whether upgraded or not cannot supply good quality water as is achieved with the deep wells with handpumps (4,5). However, rising costs, maintenance problems, etc, warrant that this simple and cheap technology be promoted to the rural populations.

Results obviously demonstrate that the upgraded well, with the improved head works, a good water raising

Table 1. Level of contamination of drinking water

WELL TYPE	NO. SAMPLED	NO. OF E.COLI/100ML	
		RANGE	MEAN
Upgraded Well	16	0 - 80	45
Improved Well			
Type 1: Wells with lining, concrete slab and windlass	8	0 - 160	75
Type 2: Wells with lining, and concrete slab (no windlass)	22	0 - 220	81
Type 3: Wells with lining, cover slab and lid	18	0 - 180	78
Type 4: Wells with lining only (no cover slab and lid)	10	10 - 280	140
Unimproved Well			
Type 1: No lining, no cover slab, with windlasses	9	10 - 400	182
Type 2: No lining, no cover slab, no windlass	6	10 - 720	263

Table 2. Range of pH and turbidity for the well waters

Well type	No of wells	Turbidity (TU)		pH	
		Range	Mean	Range	Mean
Upgraded	16	5 - 15,6	8,5	5,5 - 7,6	6,52
Lined wells (Improved)	32	5 - 26,0	21,8	5,2 - 7,9	6,58
Unlined wells (Unimproved)	32	8 - 288,0	156,0	5,0 - 7,3	6,12

system provided by the windlass complete with bucket and chain and the pit itself lined from the bottom right up to the top, delivers the best quality water - in terms of bacterial or faecal indicator load as well as the turbidity. The standard permissible level for turbidity, for community water supplies is 25 Tu and the upgraded well waters had values far below this limit. An improved headworks obviously prevents wastewater released at the surface from draining back into the well. The cover slab with lid prevents access of foreign objects, dust, dirt into the well; whilst a well lined pit prevents collapse and soil erosion, hence the low turbidity values. The latter was also apparent in this study, where the lined, improved wells had turbidity values also within the maximum permissible level as compared to the unimproved, unlined wells which had values as high as 288 TU. Turbid waters also harbour many more microorganisms in the soil particles, hence encouraging contamination.

On individual components of the shallow well, the windlass whose reason for existance is that it is there to maintain the water-raising system in a hygienic environment, i.e. keeps the chain clean and aids in raising the bucket and keeping it on a safe clean place, was shown here not to have any significant effect on the quality of water. Presence or absence of a windlass did not alter the numbers of *E. coli* in the improved wells in any way. What was apparent however, was the lining of the pit. Results demonstrate this to be most important aspect to affect water quality in shallow wells, together with the coverslab and lid, which obviously keeps all contamination out.

The findings of this study should have important application in countries like Malawi, Tanzania and various countries where the windlass is relatively unknown. For Zimbabwe, where government still provides a windlass as a subsidy, emphasis when improving a shallow well and hence water quality should be on the proper lining of the pit and the covering of the well contents.

References

Morgan, P. The bacteriology of wells - some facts and figures. Blair Research Bull: Ministry of Health, Zimbabwe, 1985.

Morgan, P. Drinking water quality for rural areas. *Blair Research Bull:* Ministry of Health, Zimbabwe, 1984.

Mcffters, G.A., Stuart, D.G. Survival of coliform bacteria in natural water: field and laboratory studies with membrane - filter chambers. *Applied Microbiol, 1972:* 24:805-811.

WHO. Guidelines for drinking water quality. Geneva: WHO. 1984.

Wright, R.C. The seasonality of bacterial quality of water in tropical developing country (Sierra Leone). *J. Hyg.* 1986: 96: 75-82.

SECTION 9

WATER SOURCES

Earth dams for RWS in Northern Region

Siaw Awuah and John Addy

RURAL WATER SUPPLY in Ghana, especially with the declaration of the International Drinking Water Supply and Sanitation Decade, has been characterised by the harnessing of groundwater for domestic and other uses. Groundwater has been exploited mainly by the use of hand dug wells and drilled bore-holes fitted with handpumps. An alternative water source being used for rural water supplies in a part of the Northern Region is surface run-off impoundment using earthdams and dug-outs.

Test drilling results and other hydrogeological conditions indicate water scarcity in the southern part of Northern Region which forms the Project area.

The area is underlain by the upper voltaian geologic formation and has these hydrogeological characteristics: (Figure 1).

1. Geologically, it is composed of impermeable rock consisting of hard sandstone and mudstones with shales (Norrip 1982). This formation does not allow effective surface water infiltration due to shallow weathering, clayey soils and lack of significant fracture patterns. Test holes show a 28% probability of success per hole (Tod, 1981).
2. Seasonal flow of streams and other surface water sources. There is a 6-month (October-April) dry period with a yearly and daily mean temperatures of 28°C and 12°C respectively. Evaporation is estimated at 1800mm for open water annually. Rainfall is relatively high during the rainy season varying from 1000 to 1300mm annually with estimated run off coefficient of 10% of total rainfall (Tod, 1981).

These characteristics imply unreliability of both ground and surface water sources in the area.

Choice of water source

Physical and socio-economic factors influenced by the hydrogeological conditions mentioned above, underlie the choice and use of earthdams and dug-outs to impound surface run-off for water supply.

The physical factors are:

1. High surface run-off during rainy season.
2. Semi-impermeable soils and bedrock, and
3. Slightly rolling to flat landscape.

High water requirement for livestock and other economic activities in the area provide the socio-economic condition favouring the use of surface water impoundment.

Between 1988-1992, the Project has assisted in constructing 11 gravity-type earthdams and 6 dug-outs to serve approximately 33,000 people and 28,500 of livestock in their design year. Total investment cost during the period is about US $5 million. (Bouman, 1992). Examples of some of the constructions undertaken during the period are presented in Table 1.

Design and construction

In collaboration with the beneficiary community, a suitable site is selected for the dam. A condition for site selection is the need for a maximum walking distance not to exceed 1.5 kilometres from the settlement though topography and soil suitability primarily determines the site.

Topographical and soil surveys are followed by design and construction of the dam and other components. The choice between a gravity dam and dug-out for a selected site is influenced by water demand, topography and availability of suitable construction materials(soils) at the proposed site. The basic components of a water supply system used by the Project are shown in Figure 2.

Main concrete spillway dimensions are determined for the once in 10 years flood with a design peak precipitation of 110mm/day/with design-peak precipitation of 165mm/day while the once in 100 years flood is used for the dimensions of the auxiliary earth spillways.

Construction is undertaken by the Project with its earthmoving equipment. This follows a sequence of activities which are:

Access road preparation; setting up of camp for workers; etc.

On average, 4 months are used in building a gravity dam and 2 for a dug-out. Construction is normally undertaken in the dry months of October to May.

Implementation of project

Design of the water supply system is done solely by the Project while site selection and construction is jointly undertaken with the beneficiary community. Land and cash contribution equivalent to 10% of total cost of scheme, are made by the community prior to start of construction. Food and labour contribution is made on daily basis by the community during construction. Construction equipment and materials are provided by the Project.

A two-year guarantee period excluding construction period is given for the scheme by the Project during which all remedial works on any component of the water supply system are carried out at no cost to the beneficiary. Training of village maintenance team members is done during this period.

Water sufficiency

Yield and drawdown characteristics of some of the reservoirs indicate availability of water in sufficient qualities throughout the year. Table 2 is a presentation of the yield and drawdown characteristics of some of the reservoirs.

Constraints and problems

Notwithstanding sufficient supply of water made possible by the use of earthdams in the Project area, some pertinent problems are encountered which undermine Project efforts. High evaporation and seepage rates, unreliable intake system, excess water demand and poor water quality are the critical problems.

The evaporation and seepage criteria used seem an underestimation in the design of some reservoirs thus contributing to excessive water losses. The Kukuo reservoir, designed and constructed to similar standards as others, has since construction in 1991 been getting dried by February each year.

A floating intake system consisting of a 75 or 50mm high density polyethylene pipe(HDPE) fitted with a strainer(made of perforated HDPE) and floater is used. Plastic containers (jerry cans), widely available in the markets, were used as floaters for the intake. Cracks develop on the plastic containers with time and result in the sinking of the intake to the reservoir bottom (floor).

The actual population served by a dam is difficult to forecast due to use by neighbouring villages which are either as near as a kilometre, or 5 or more kilometres far from the dam. The need for user/beneficiary contribution to and participation in the construction of the scheme disqualifies distant villages from being part of a water supply scheme while unwillingness to join in development efforts with other villages prevent nearby villages from being part of the scheme. However, at the peak of the dry period, all neighbouring villages depend on the reservoir creating excess demand for water.

Like any surface water source, pollution of reservoirs is considerably high in the Project area. Poor sanitation and agricultural practises in the catchment contributes immensely to water pollution in the reservoirs.

Nitrates and other inorganic constituents though not monitored, are likely to exceed acceptable limits in view of the use of inorganic fertilizers by farmers in the catchments. Observed periodic algal blooms in most of the reservoirs presupposes the presence of these nutrients.

Siltation in reservoirs is not monitored. However, intense crop cultivation coupled with loose top soils in the Project area obviously contribute large quantities of sediments to the reservoirs.

Interventions

Effort is made by the Project at finding feasible solutions to the constraints mentioned above.

1. Insufficient soil investigation, based on the assumption of geologic uniformity in the Project area, is basically responsible for excessive seepage loses in some reservoirs. Village labour is used in making test pits for subsurface investigations. Pits often terminate at the hard laterite layer which in most cases is a meter below the surface. Digging tools used cannot penetrate this laterite layer and thus make it impossible to identify soil below the laterite layer. A drill has been acquired by the Project for subsurface investigations.
2. Low density plastic containers have been replaced by a high density and more UV-light resistant plastic floaters. The new floaters have been used for 11 months without being damaged compared with the average of 6 months that the low density containers were used.
3. An excess reservoir capacity is to be added to cater for future dependent villages not catered for in the design;
4. Direct contact with water in the reservoir meant for human use is avoided by the use of collecting wells and fence.

Monitoring on guinea-worm infestation in the reservoirs by the Danish Bilharziosis Laboratory in Tamale over a one year period showed no infected cyclops(intermediate host) in any of the Projects reservoirs (Kees, 1992).

However, the general incidence of guinea worm increased by 34% between 1989 and 1991 in the district where the reservoirs are situated (Bouman, 1992).

Sloping sand filter and an infiltration gallery are used in two of the dams for water treatment. The infiltration gallery reduces water turbidity from 200NTU to 5 NTU. The extent to which these filter systems improve bacteriological quality is difficult to assess due to possibility of external contamination of wells by drawing buckets etc.

Conclusion

Hydrogeological and socio-economic factors impose a limitation on the use of any alternative water supply system in the Project area of Village Water Reservoirs. The high cost of the water lsupply schemes, the complex technology used in designing and construction and associated constraints indicate a tremendous challenge to the sustainability of the water supply schemes at Project and village levels. Similar challenges can be expected elsewhere where physical and social factors dictate the choice of surface water as the only feasible alternative. Research is therefore needed on low-cost surface water supply systems for areas of similar physical and social characteristics if the objective of the International water supply and sanitation decade are to be met in such areas.

Table 1. Examples of constructions undertaken since 1988

Community	Type of construction	Year of construction	Design population Human	Design population Live-stock	Capacity at full supply level (m^3)	Cost in cedis	Cost per capita (Human) (in cedis)
Chirifoyili	Gravity dam	1988-89	6930	4500	135,650	35,181,261	5,077.00
Buyili	Dug-out	1989-90	360	400	13,500	11,327,665	31,466.00
Aseiyili	Dug-out	1989-90	300	365	10,000	6,200,334	20,662.00
Yong-Dakp.	Gravity dam	1989-90	2150	1800	81,000	16,552,840	7,700.00

Table 2. Yield and drawdown characteristics of some reservoirs

Reservoir	Type of construction	Capacity (m^3)	Designed Full supply level (m)	Designed Volume (m^3)	Designed Lowest supply level (m)	Designed Volume (m^3)	Observed 1990 Highest Level (m)	Observed 1990 Lowest Level (m)	Observed 1991 Highest Level (m)	Observed 1991 Lowest Level (m)
CHIRIFOYILI	Gravity Dam	135,650	9.77	135,650	6.21	5,025	9.71	6.18	9.76	6.16
BUYILI	Dug-Out	13,500	9.80	13,500	6.55	2,300	9.78	6.50	9.79	6.50
ASEIYILI	Dug-Out	10,200	8.60	10,000	6.63	3,200	8.55	6.60	8.56	6.59
YONG	Gravity-Dam	31,000	9.50	81,000	6.67	2,500	8.41	6.65	9.39	6.52

References

BOUMAN, D. (1992): Water supply by dam construction on Village Scale: the experience of the Village Water Reservoirs Project in the Northern Region of Ghana. 1st draft (Unpublished)

KEES, H (1991): Water quality in Village Water Reservoirs; with and without filtersystem as constructed by Village Water Reservoirs Project in the Northern Region in Ghana. (Unpublished).

NORRIP (1982): "Phase 1, Test Drilling Report" Norrip Water Sector, Tamale, Ghana (Unpublished)

TOD J (1981): Groundwater Resources in the NorthernRegion, Ghana. Norrip sectoral Report (Unpublished).

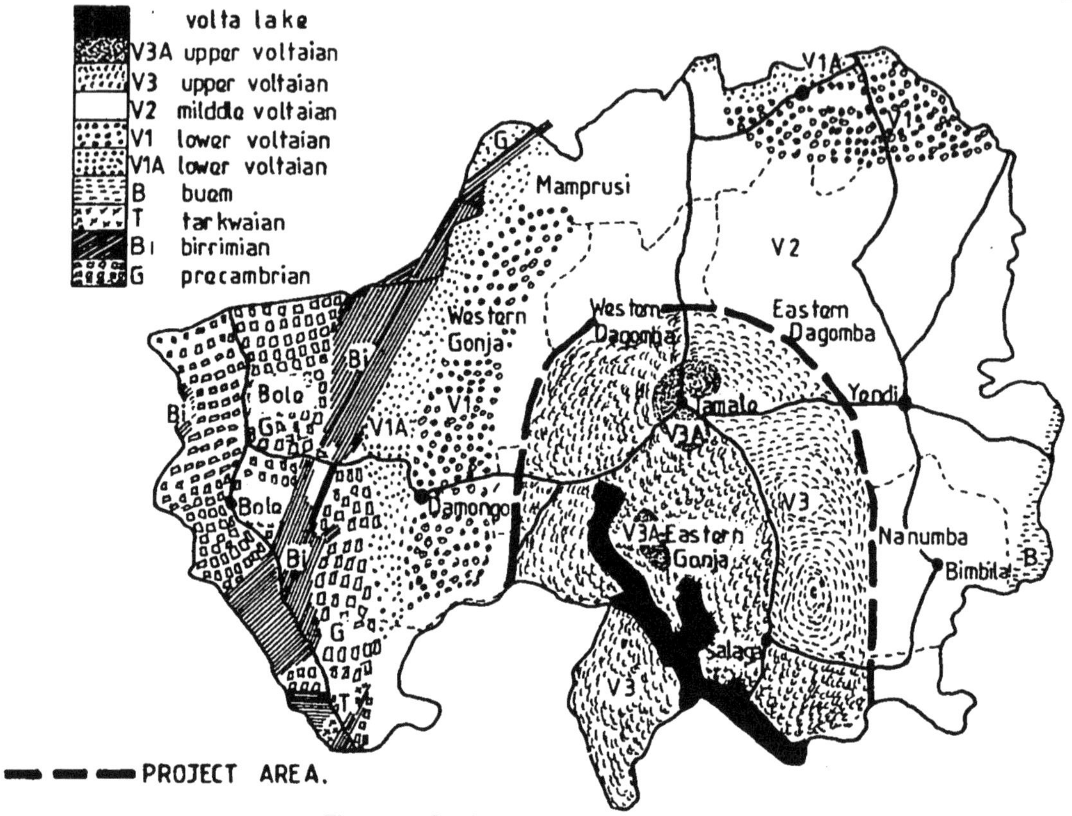

Figure 1. Geology of Northern Region
(Source: Tod, 1981)

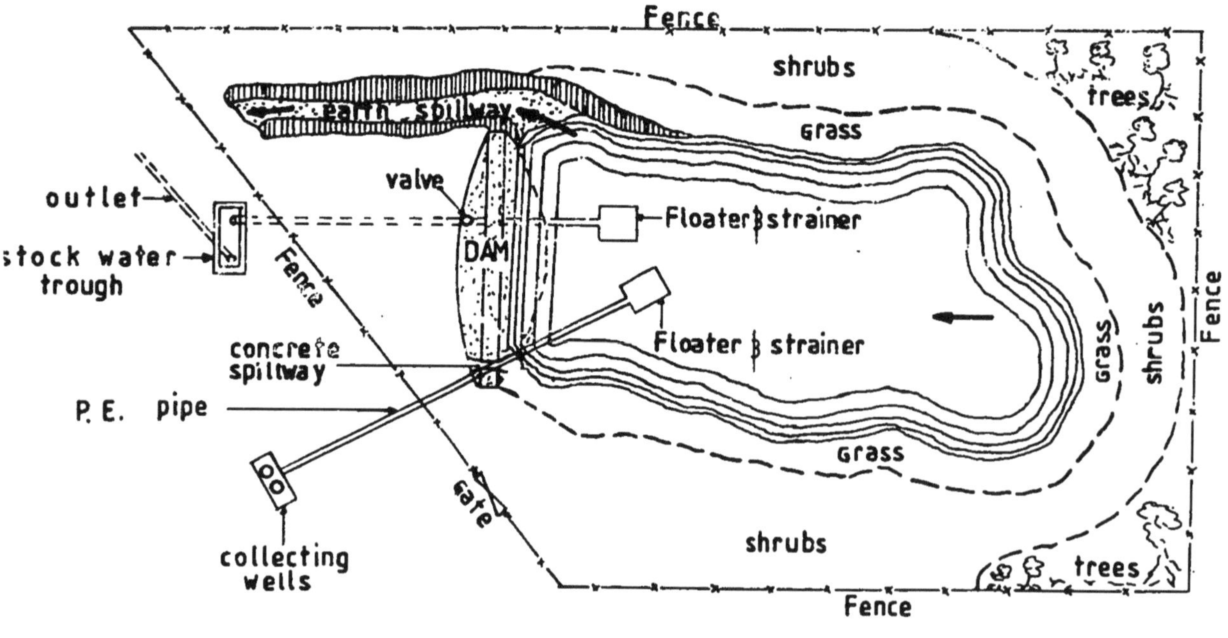

Figure 2. A typical layout of a water supply system

Rainwater harvesting initiatives in Ekpoma, Nigeria

Layi Egunjobi

IN 1989, I had an assignment as part-time consultant to the Federal Ministry of Health to monitor and evaluate primary health care activities in Bendel State in Nigeria (Bendel has since been split into Edo and Delta states). During the three-month period of the assignment, I noticed a widespread use of simple structures attached to houses to collect rainwater for domestic use in Ekpoma and other settlements around. My curiosity in this respect led to follow-up visits to this town making further observations and asking question from a cross-section of community members. The result of this simple exercise is presented in this short paper.

Background

Ekpoma is the administrative headquarters of Esan West Local Government Area in Edo State. With an estimated 1992 population of 63,467, the town is made up of twenty-two traditional wards (clans). It is located 80km north of Benin City which is the state capital. Ekpoma is located on the Ishan Plateau. Its characteristic physical features include a level topography, loose sandy soil and the paucity of surface drainage. There is a marked dry season the duration of which may extend to five months. The mean annual rainfall in the area is 1,556mm. The soils are porous with the result that the water table is very low. This explains the seasonal character of the streams which dry up during the dry months. The natural vegetation is deciduous forest which is very rich in timber resources.

One of the major socio-economic problems facing the area is shortage of water supply. As can be inferred from the foregoing description, the natural characteristics of the area are not favourable to abundant supply of water. Secondly, public sector intervention in the provision of water has been rather slow and largely ineffective. For instance, it is only very recently that a public bore-hole was installed within the settlement. At the time of writing this paper the bore-hole did not function as it was reported as having collapsed. The second borehole which was privately owned remained the only communal source of water. The cost of water from this private source was unaffordable to a majority of community members. The only readily available sources of water in Ekpoma therefore has constantly been the rainwater harvesting gadgets attached to dwelling units. This is an initiative that has been at work with regard to the satisfaction of the people's survival instinct.

The rainwater harvesting gadget

The idea of rainwater harvesting in Ekpoma and the surrounding settlements rests on the collection of rainwater that falls on the roofs of houses and channelling the water into an underground storage tank or reservoir. The main elements of the system are the roof-gutter, the pipe and storage tank (which community members refer to as 'wells'). With respect to a number of systems observed, there are additional fittings such as funnels with wire gauze.

The operation of the system consists of harnessing the rainwater that falls on house roofs. Most of the roofs in Ekpoma were made of corrugated iron sheets. The entire roof-edges or parts are fitted with gutters made of iron sheets. A pipe cast from iron sheets is fitted to the end of the roof-gutter and directed into a small opening leading into the storage tank. The tank is made of concrete blocks deep down the ground with part of it elevated above the ground level. Of course there is another opening of about 0.6m by 0.6m through which water is drawn using small buckets (steel or plastic) that is tied to a strong rope.

As important aspect of the gadget is the size of the storage tank; the bigger the tank, the more the volume of the water it can hold and the more the number of people the water will serve. There are variations not only in sizes of tanks but also in their shapes. Most of those observed were circular while others were rectangular or square. With regard to the circular tanks, the average size is 2.5m in diameter and 3.5m in depth. This works out to a capacity of 17,899 litres of water when the tank is full.

Apart from the shapes and sizes, there are also variations in the types of finishing. While quite a number of the water systems were covered with corrugated iron-sheets, a few were covered with concrete slabs; yet others were left open with poles or planks loosely stretched across the opening. Community observations and discussions with members indicated that variations in the finishing are a function of income. In fact a very minute proportion of households already fitted overhead tanks as part of the system; water is lifted from the underground concrete tank into the overhead steel tank by electric water-pump, and thus distributed into the house by gravity.

Cost of construction

The materials used in the rainwater harvesting systems are sand, gravel, cement, iron rods (as applicable) and planks (as applicable). The types of material vary directly with the types of storage tanks desired. A tank with a concrete slab for example, does not require iron sheets. By the same token, the quantity of materials is in direct proportion to the size of the tanks.

An average sized tank of about 18m3 will cost N13,709 ($US 343) going by 1993 prices. The same system cost only N1,169 ($US 29) five years ago. This reveals an astronomical rise in prices in the country during the period. Nigeria, it will be recalled, adopted the structural adjustment programme in 1986. The naira value has since fallen from $1,00 to N1.00 in 1985 to $1.00 to N40.00 in 1993. With this escalating rates in the prices of goods and services, lack of affordability has constituted one of the most formidable constraints facing construction of new and maintenance of the old systems.

Operation and maintenance

The quantity of water an average sized tank will hold at full capacity is about 18,000 litres. When installed in a dwelling unit of two households (national average household size is 6.8), the storage tank water will last for 88.2 days or approximately three months at a minimum consumption rate of 15 litres per capita per day (a WHO survey indicated 15 litres as minimum average daily consumption for rural areas of Africa). However, it should be noted that water drawn out of the tank is replenished by subsequent raining during the rainy season – i.e. May to October. It is during the dry season that water consumed from the reservoir is not replenished. This in effect is the period of critical shortage when the harvested rainwater is most useful to community members.

As to the quality of rainwater harvested, it can be observed that the water is generally not potable even though it is used multipurposely (including drinking). No laboratory tests have been carried out; however, discussions with health workers around revealed that harvested rainwater was generally unsafe for drinking without some form of treatment. The reason, it was gathered, has to do with storage and use of the water. The first few rains in the year inevitably collect a lot of dust not only from the atmosphere but also from the roofs where dust might have collected in large quantity during the dry season. It was observed that while some households turn the collecting pipes away from the reservoir during the period to prevent impure water entering, others did not. Another factor contributing to the presence of impurities in the harvested rainwater is inadequate cleaning of the tank. Most tanks are washed invariably only once during the dry months when the water level falls almost to the tank floor as families will not devote sufficient quantity of scarce water resources for this purpose.

Discussion

The origin of rainwater harvesting initiative in Ekpoma area has not been ascertained. Discussions with a cross-section of community members shown that the first such system was probably constructed by the early missionaries who worked in St. Peter and Paul Seminary in the town. Others claimed that the practice was as old as the community itself. However, the fact remains that it is a widespread practice with one in every four or five houses having a rainwater harvesting system. The practice has since become a survival strategy as they are about the only water sources available between November and March. When tank water is completely used up, community members have to travel up to six km to fetch impure stream water that is found in small ponds.

Table 1
Cost of construction of rainwater harvesting gadgets (1988 and 1993)

Material/Labour	Quantity	Rate (Per Unit) (N)		Cost (N)	
		1988	1993	1988	1993
Materials					
Sand	1.5 lorry loads	120	350	180	525
Gravel	1 lorry load	250	750	250	750
Blocks	384 No.(9"x9"x18")	2.5	10	960	3,840
Cement	18 bags	40	168	720	3,024
Iron - Sheet	1 Bundle of 20 sheets	32	1,400	32	1,400
Planks	10 No (2"x3"x12")	1.5	25	15	250
Nails	816	1.5	10	12	80
Labour					
Digging	20 man-days	15	80	300	1,600
Bricklaying	20 man-days	15	80	300	1,600
Carpentry	8 man-days	10	80	80	640
Total			-	1,169	13,709

The initial efforts in the construction and installation of rainwater harvesting systems were crude; however, with time incremental refinements were added to the gadgets to increase their utility value. As mentioned earlier, some members of the community including those in the Seminary have added electric pumps and overhead tanks to the systems.

The problem facing potential rainwater harvesting system owners is the currently high prices of building materials (as prices in other sectors of the economy).

A second problem observed is technical incompetence in the design and construction of a number of the gadgets. Some concrete walls of the storage tanks were collapsing while some tank roofs were caving in and therefore constituting danger in the premises.

A third problem relates to factors that detract from the safety of water with regard to drinking. For example, it was observed that where storage tanks were covered with concrete slabs, these were used by household members in drying food stuff part of which sometimes escaped into the tank. It was also observed that roaming domestic animals (e.g. goats, sheep, dogs) used the slab tops as places of abode on which they defecate.

Recommendations

In view of the foregoing descriptions and discussions, the following recommendations are important in strengthening the people's initiative and therefore making rainwater harvesting system more affordable and effective:

- Public sector intervention should include encouraging people or compelling them to integrate rain water harvesting with the overall housing plans submitted to Town Planning Authorities for approval. The construction of the system thereby goes on simultaneously with the construction of the main building.

- The operators and users of rainwater harvesting systems need to be given required technical assistance in construction and installation. A relevant government agency should establish a technical assistance unit and identify communication channels between the unit and the public to make this possible.

- Community members through People's Bank and Community Banks should be made more easily accessible to soft loans specifically for construction rain harvesting gadgets.

- Relevant government agencies should embark on health education programmes that emphasize aspects of sanitation and personal hygiene with reference to the use of water from the rainwater harvesting systems.

SECTION 10

WATER SUPPLY

Pumps, people and payments

Dr Manu N Kulkarni

BETUL DISTRICT is one of the forward-looking districts of Madhya Pradesh, where NGO innovations and Partnership programmes with the District administration have always been in the forefront of development actions. The District is the centre of Evangelical Luthern Church activities like Hospital Care Schools and Water Development Activities. A Hand Pump demonstration project was initiated in late 1987 by Government of M.P and UNICEF to assess the reliability and maintainability of India Mark III (VLOM) and improved India Mark II. Twenty Five handpumps of each type were installed and monitored and was found that India Mark III (VLOM) has a meantime failure of 18 months and almost 90% of the repairs could be carried out by a village level mechanic. This was the prompting factor for taking up large scale awareness and training for VLOM and sanitation interventions with communities spearheading the programme since early 1992. Betul was one of the 10 development Blocks of Betul District which was selected for this experiment. "Successful Community Water Supply Programmes (CWS) involve a combination of hardware and software technology and institutional/ organizational support elements matched in such a way that each community recognizes the benefits of improved supply and can afford at least the costs of operating and maintaining it and has the skills, spare parts, materials and tools available to sustain it. To maximise health benefits, parallel investments in health education and sanitation programmes should be planned alongside CWS improvements" (World Bank, 1987). As one of the activities in support of International Drinking Water Supply and Sanitation decade, VLOM was promoted as means of overcoming some of the major obstacles to sustainable water supply systems. Many past failures of CWS systems can be blamed on the inadequacies of central maintenance in which a Water authority despatches teams of skilled mechanics with motor vehicles from a base camp, often serving a large District, to respond to requests.

Response to handpump breakdowns

At present in Betul District there are about 5000 India Mark II Handpumps and about 120 Mark III. Each Block level mechanic is servicing about 400 Handpumps and it is physically not possible to attend to their breakdowns. The present system of informing the breakdown by the villagers is also not efficient.

Any villager can inform the Block level mechanic by writing the complaint in the Register kept in the Block.

In some Districts of M.P. some villagers telephone the Executive Engineers about the breakdown. The Block level mechanic keep looking at the register of complaints and decide to "attend" to them. In some places the weekly markets are used by the villagers to inform the breakdowns, where complaints are collected by the PHED. Thus there are different ways of informing the breakdowns by the villagers. This haphazard information system does not ensure quick and efficient response.

Villagers who are ignorant of the nature of the breakdown simply magnify the breakdown, even if the breakdown is minor. This is where the peoples knowledge of the Handpump becomes important. Given this present situation of Hand Pump functioning in Betul the process of building community awareness on the entire range of Water and Sanitation issues assumed urgency. Thus emerged the WESA (Water and Environmental Sanitation Awareness) concept in this experiment.

Demand generation through WESA

WESA was spearheaded by the NGO in Betul with the objective of:

(a) Building community awareness on water and environmental sanitation.
(b) Forming WATSAN committees in each of the 160 Project villages, involving 120,000 population.
(c) Selecting by the people of the village their own Hand Pump User Representatives, who will select village handpump mechanic team with two women and a man.
(d) Train the selected Hand Pump Mechanics in the operation, repair and maintenance of India Mark II and III (VLOM) pumps and help the Public Health Engineering Department in the conversion of Mark II pumps to Mark III and their subsequent maintenance.
(e) Building the village WATSAN fund for upkeep of Pumps and sanitation facilities like toilets, soakage pits, drainage, cattle troughs etc.
(f) Ensure the safe custody of Handpump spares and tool kits.

WESA process has created in the past months a tremendous sense of community participation for WATSAN programme. (Village WATSAN Committees formed - 172; Orientation of WATSAN Committee members - 1215; Orientation of HP Users Representatives - 536, Women mechanics trained - 116, men mechanics trained - 44). The process has helped to evolve women as the principal actors in this whole experiment. The Village elderly men have come to believe that Water and Women go together because women's life is linked to water lifting, hauling, storing and using efficiently or inefficiently. One Village head man described such women as "Chaturis" (Intelligent). On the other hand village youth, though smart, cannot stick on to the village and even if trained, they would demand service charges.

Hence the selection of married women of the village for Hand Pump training emerged, married because they do not leave the village and go away and are available all 24 hours and at least part of the whole year! This helps the repair and maintenance of VLOM pump better.

WESA process highlighted in their awareness camps, use of Water, better handling of Hand Pumps, better drainage around the pump and platform, use of soakage pits, cattle troughs, use of toilets, and better disposal of waste. In some Camps women asked questions on use of bio-gas, improved chulas (stoves). The training was participatory and village men sat behind and enjoyed their women folk asking questions. The whole village had a festival look and the flirting banners on safe water, posters on Hand Pump, wall paintings, videos and audios - all contributed to the effective communication of WATSAN messages. The video -Stree Shakti - (Women power) depicting how women can repair Hand Pumps was screened and many women got inspired to try their hands.

Focus on women as pump mechanics

Early experiments in India on selection of Hand Pump mistries (mostly young boys repairing cycles, electric motors etc) as Hand Pump repairers and mechanics did not involve women in the entire range of WATSAN activities but confined only to repair of hand pumps. "Tilonia started experimenting to find out if it was possible for semi literate boys to repair the 300 handpumps the SWRC had installed. We trained them initially for a month. We found it just required some manual labour and common sense (Sanjit Roy, 1989)".

WESA's approach has been therefore different and took into account the "parallel investments" in health education and sanitation programs along side CWS improvements. As the WESA covered more villages, more requests came in for providing help to build pour flush toilets and improve the village drains. They found just in $80 dry pit pour flush latrines can be built and the repair of the pumps does not need `secret knowledge' and the villagers who were exposed to WESA orientation came forward with money to demand more knowledge and action to build toilets. Some villagers collected anything between $10 to $412 from the interested villagers to defray the cost of building individual toilets and the cost of repair to Hand Pumps. They were not clear about the cost but their 'enthusiasm' was reflected in their contribution. Some villagers did not keep the WATSAN fund in Banks but loaned the amount to the needy on interest but with the assurance that when the money is needed they will return it to the treasurer, who was in some villages, a barber! Why a Barber because the barber would always have cash and will not run away from the village.

VLOM training

As the demand generation on WATSAN caught up there were more requests from Village WATSAN committees to organise the training on Hand Pumps. Each Village WATSAN Committee nominated two village women and a man to undergo VLOM Training. When this training was launched two problems were faced (a) The existing Block mechanics of the PHED were little sceptical of such a training for village women. How can these semi literate women handle such a pump? Without the cooperation of Block mechanics such a training cannot sustain because they have to cooperate with these women and the women have to cooperate with Block mechanics to ensure functioning of pumps in the villages. There was also the fear of the unknown - these women would replace us, where will we go? Hence joint consultations were held with the Block mechanics under the leadership of the dynamic Executive Engineer of PHED of Betul District.

The NGO had already surveyed the Mark II Pumps which were not functioning and the Mark III (VLOM) pumps which were to be installed (converted from Mark II). The PHED had furnished the list of both types of Pump. The plan is to convert all Mark II pumps into VLOM (MarkIII). The training was hands-on. As the training progressed the women demystified the English names of the parts of the pump. The following local names were given by these women to several parts of the Mark III pumps.

1. Plunger Rod = Chotu Chad.
2. Plunger Yoke body = Pinjara (Cage)
3. Upper Valve = Mendki (Frog)
4. Washer = Bucket
5. Spacer = Katori
6. Follower = Glassi (Like a glass)
7. Check Valve body = Badi Pinjara
8. Check Valve = Badi Mendki (Big Frog)
9. Push Rod = Choti kila kanta.
10. Cylinder Body = Pyle (measuring grain)
11. Inspection Cover bolt = Tala
12. Inspection Cover = Dhakan
13. Double end spanner = 19 & 17 Chawi
14. Head = Peti.

The aim of VLOM concept has been:
Pumps are easily maintained by a village care taker, requiring minimum skills and few tool.

- Manufactured in-country, primarily to ensure the availability of spare parts
- Community choice of when to service pumps.
- Community choice of who will service pumps
- Direct payment to the repairers by the community (World Bank, 1987, Page 13.)

In the extended VLOM concept M means management of maintenance.

The VLOM training in Betul recognised this M factor from the beginning and the training style was oriented to drive home the M factor all the time with the trainees. In Betul District the PHED through their Block Mechanics had adopted the breakdown repair as the most common form of maintenance. As a result of this training large number of Hand pump mechanics, nominated by the village WATSAN committees, came to know about the preventive maintenance, like simple greasing of the chain, tightening of nuts and bolts, which has helped to spot the problem in good time.

Assessment of the VLOM training

How do we assess the knowledge of these women/men trainees in VLOM pump? The NGO in consultation with the PHED Executive Engineer decided to provide the Spares and tool boxes in joint custody of the village Panchayat and Village WATSAN Committee, and these were to be used by the trained mechanics of that village.

We adopted a simple test. We spread all the parts of India Mark III pump on a table. We called each of the trained team (Two women and a man). We picked some part of the VLOM and asked the women to join them in 3 minutes. They did it with perfection. On one occasion we mixed up the parts of Mark II and Mark III (VLOM) and asked them to join. They could tell us they are separate and cannot be joined! Such was their knowledge of VLOM parts and the whole.

The future

There are some constraints in the whole experiment. The conversion of Mark II into Mark III cannot be done in all cases since the Inside Diameter of borewell/tubewell fitted with IM II pumps do not match with the Mark III pumps. Spare parts and tool kits have to be properly distributed to all the trainees so that they feel that they are "empowered". The partnership between State PHED, NGO and UNICEF has to be strengthened in order to sustain the future VLOM strategy in the remaining blocks of Betul District. Parallel investments in sanitation activities have to be stepped up to have better mileage for the communities in WATSAN tasks.

Reference

A) Charles Kerr, *Community Water Development*, Intermediate Technology Publications, 1989. See Sanjit Roy, Page 179
B) Saul Arlosoroff et al, *Community Water Supply: The Handpump Option*, World Bank - 1987

The author is thankful to Mr S. A. Wahid, Mr Mansoor Ali and Mr V. R. P. Nair of UNICEF, Mr Satish Raghu, WESA Betul, Mr Tawani, Executive Engineer of PHED - for their valuable suggestions in the preparation of this paper. However, the author alone is responsible for any omissions.

Pipeline extensions spread benefits

Peter Smith and A Mbaye

MANY OF THE water supply schemes built in Senegal in recent years are operating below capacity (Smith, 1991), partly due to poorly developed distribution systems. In a small scale, ODA funded project, an extra 25,000 rural people have been connected to existing supply schemes by construction of a large number of short extensions from the original distribution networks.

Background

Senegal was severely affected by regional droughts in the 1970's and early 80's, and even in normal years there is low annual rainfall, (typically 200 - 700 mm), and a long dry season. The country is very flat, and groundwater is the principal source of water. During the droughts many shallow wells dried up, and so a policy decision to develop deep, motor driven, pumped boreholes was made. With the support of a range of donors and funding institutions, over 500 new such schemes have been constructed since 1980.

Among these schemes were 18 which were funded by a subvention from Britain and constructed during 1985-87 (Horsfield, 1988). These were typical of many projects, each comprising a borehole 100 - 300m deep, with a vertical shaft pump driven by a diesel engine. Water is pumped to a 100 - 200 cu m water tower, and supplied to villages up to 6 km away by gravity pressure, through a number of pipelines radiating from the water tower.

It did not prove possible to complete all the intended distribution lines from the water towers during the project period. Following requests from the un-served villages for the pipelines to be built, the first phase of a programme of self-help extensions to the original schemes was carried out in 1989/90. Further phases were undertaken in subsequent construction seasons, and the fourth phase has just been completed, bringing the number of extensions built to 50, totalling about 100 km. The first two phases of construction were entirely at boreholes funded by ODA in the earlier project. The third phase included several schemes originally funded by other donor institutions, and all fourth phase extensions are at 'non-ODA' schemes. A fifth phase is planned.

Experience gained each year has led to developments and improvements in the way that the extensions are implemented. The intention of this paper is to illustrate the working methods employed, and to bring out aspects which have contributed to the success of the project.

Funding has been on a limited scale, more often associated with an NGO than a major bilateral donor. This has highlighted the need for efficient working practices, to minimise costs, and to maximise the number of beneficiaries.

The time taken to discuss and approve proposals has been reduced as a result of the good working relations developed between the various parties over a period of years. Bureaucratic procedures sometimes associated with donor funded projects have been kept to a minimum. This has enabled a flexible approach to be adopted, with rapid response times at all stages.

Getting going

Agreement in principle to fund an extensions programme has to be reached by ODA, and a firm programme within this sum can then be developed and proposed.

As a starting-off point for an extension to be considered, a request for an improved water supply has to be received from the village, with some indication of their willingness to be involved in its construction.

A preliminary assessment of the viability of a request is carried out at the Rural Water Department HQ. Data are available on the existing borehole schemes which would serve the villages, including populations, type and capacity of pumps and water tower, distances etc. This allows an initial 'weeding out' of obviously unsuitable proposals. The basis used in the first place to assess proposals is the cost per capita. As a guideline, a maximum cost of about £25/person is looked for, and as a rule of thumb, this translates to a population of 200 at a distance of 2 km, 300 at 3 km, etc.

Village assessment

A visit is made to each of the potential villages in order to verify technical and social aspects of the proposal, and to cross-check the data available in the office. Points which are given attention during this visit include:-

Feasibility. To ensure adequate pressure will be available at the village to be served, the existing water tower height, and intervening terrain have to be checked. In the past, different projects have used varying design criteria, and the water tower height may be too low. Water towers should generally be 15 m high, and the distance to the village less than 5 km. The very low relief topography of Senegal means that a full survey is not normally required. Distances involved are measured to the nearest 100m using the vehicle speedometer. This information helps to determine whether an extension or some other form of water supply is the most appropriate.

Spare capacity. The extensions are intended to make use of underused boreholes, where existing use of the motor is less than say 4 - 5 hours/day. The operators log book is inspected to find out the current and recent useage.

Mechanical condition. The pump and motor should be in good working order and have a reasonably trouble-free recent history, also seen from the log book. The general appearance and condition of the equipment is a good guide to the interest and capability of the operator, and to the likely future operating record.

Village population. A very approximate check on the village size can be made on the basis of the number of houses. This acts as a cross check on the possibility of recent population changes and guards against census errors.

Other Parties. Water-using projects such as agro-forestry or cattle developments may be under preparation. Information on such projects is sought in discussion with the village headman who should be aware of such plans. At the same time the possibility that other funding agencies may be planning a well or other water supply can be looked into. A visit is also made to the local administrative officer to check on these matters and advise of the project. It may be very helpful to consult religious or other community leaders who can assist in community organisation.

Village co-operation. Villagers will be expected to host the artisan labour and to dig trenches to their village without pay. A general meeting is held with the population to confirm this aspect. At the same time the number and location of standpipes can be agreed, with villagers taking the lead in this decision.

Inter-village co-operation. Villagers will manage the scheme on completion, and the long-term success of an extension depends on there being a harmonious relationship between the villages to be linked. The existing water committee at the borehole must agree with the idea of the extension, and if possible the new village should be represented on the committee. There should be no ethnic difficulties between the villages.

Design

Design is kept simple and standard wherever possible. Trenches are dug 350 mm wide and 800 mm deep. Pipes, in the range 63 -150 mm, are sized on the basis of informed experience depending on distance, population, and topography. The extensions are normally intended to provide for domestic consumption only.

A robust and economic 4-tap standpipe design - a square pillar 1.20 m high by 0.4 m set in a 3.0 m square apron - was used. Each needs about 2 cu m (5 tons) of concrete, reinforced with weldmesh. Cement is purchased in nearby towns, and sand and aggregates obtained from local sources. Graded laterite is used for aggregate, and may have to be transported some distance by horse and cart.

Four locally obtained taps are fitted, and although these may be of lower quality than specially designed imported models, they are more readily available when the time comes for replacement. A perimeter drain leads waste water to a nearby soakaway. As a guide, one standpipe per 200 - 250 people is built, but this may be varied to suit village structure and customs.

Project preparation

Once a potential extension has been assessed on the ground and the pipe lengths, diameters, etc. determined, the estimated cost of each can be worked out. All viable extensions can be ranked on a cost/capita basis to help determine the most effective use of funds. A proposed programme of work is drawn up by the Ministry Engineer and ODA co-ordinator.

This information is incorporated in a project framework document, and submitted for confirmation of funding. Recent phases have benefitted from past experience, reducing delay at this stage. Where funding has been previously agreed in principle, extensions assessed in October have been started December, and completed by February.

Implementation

An implementation programme is drawn up and pipework deliveries are co-ordinated with the supplier to suit this. Galvanized pipework for the standpipes is pre-assembled as far as possible. All pipework is delivered to site by the supplier. Several sets of re-useable timber shutters are made up for the standpipes.

Villagers are provided with hand tools to do the unskilled work - excavating, pipelaying, and backfilling trenches. A small group of artisans comprising a joiner and a mason (who concentrate on the standpipe construction), and a plumber and foreman to supervise PVC pipelaying and jointing and to set up the standpipe pipework travel to the village. This group moves from site to site, and are paid for their work.

Progress of trenching in sandy soils may be as rapid as 2 km/week. Trenching is generally done by young men from the villages, and women and children are also sometimes involved. Standpipes are built at the rate of one every six days or so, again with village assistance to grade aggregates, mix cement, transport materials etc.

At each village, a supervisory visit by the Ministry Engineer is made at the start and finish of construction, and at least one intermediate visit. He is available to sort out any problems which may arise, as required, and may draw on Ministry equipment eg transport.

Financial

The principal costs are for supply of pipes and materials, and artisan labour for the standpipes. The cost of each standpipe including labour and materials is about £400. Pipes vary in price from £1.7/m for 63 mm dia to £6 for 150 mm dia. In overall terms, a budget of around £120,000 allows construction of 35 - 40 standpipes and 35 km of extensions, which can be completed in a 9 month working period. This would provide piped water to around 7,000 people in say 15 villages, at a cost per capita of around £17. A typical breakdown of expenditure is as below.

Pipework	84%
Artisan wages	6%
Standpipe materials	6%
Fuel, Supervision, Tools, etc	4%

Benefits

The obvious benefit is to the village served by piped water for the first time. Follow up visits have indicated that a range of benefits do in fact materialize. Incidence of gastro-enteric disease is reported to drop. Women have more time for productive activities and marketing. Commercial activities such as vegetable growing begin. Some villages where population was in decline due to water problems have seen this reversed, and populations have increased again. The involvement of the community in the construction has lead to a better stewardship of the water supply once in regular use. The success of the completed schemes has encouraged other villages to seek assistance in the same way.

There are also gains resulting from the increased use of a borehole by more people. Low cost extensions are far more cost-effective than complete new schemes, and also cheaper than wells and handpumps in most cases. Better utilization of existing infrastructure means that the return on the original investment, whether paid for by donor or by loan to the Government, is improved. A larger population can more easily support the running costs of the borehole, and is more likely to make good use of the installation.

Conclusion

The project successfully combines the particular strengths that each party involved can contribute to it. The active involvement and co-operation of the communities, the Ministry, and the donor have resulted in an efficient and effective programme. A large number of people have been served with improved water at low cost.

There is considerable potential for this type of project, and in particular for funding agencies to take on such work at schemes originally funded by others to the overall benefit of the population. The success in Senegal has encouraged the Government to seek larger scale funding from other agencies for a much larger programme of extensions.

References

1. HORSFIELD A. *A Senegal Village Water Supply Project.* J.I.W.E.M. 1988 2 (4)

2. SMITH P G S. *Towards More Effective Projects.* 17th WEDC Conference, Nairobi, 1991

SECTION 11

WATER TREATMENT

Chlorinating household water in The Gambia

Dr Christopher J Austin

THIS STUDY INVESTIGATED the feasibility of providing safe drinking and cooking water by in-home chlorination of household water jars (HWJ). Open well water quality was measured in the dry and rainy season and found acceptable for chlorination. Rural village women were taught to dose their HWJs with a diluted household bleach solution to give an organoleptically acceptable dose of 2.0 mg/L chlorine. This dose eliminated fecal coliforms within 30 minutes and protected stored water for 24 hours. A 22 village double blind randomized intervention trial was carried out over a rainy season. No effect of HWJ chlorination on the incidence of diarrhea was detected. A trend was revealed that for children 6-24 months, villages which chlorinated the HWJs did not suffer a significant decrease in the village mean weight-for-height Z-score as compared to control villages (p =.1170). In children 6-24 months and also 25-60 months, the control group suffered a significant increase in the proportion of malnourished children (p=.0002); whereas the intervention group did not experience a significant increase (p=.1000). The study concluded that women chlorinating their HWJs may be an appropriate avenue of providing safe drinking and cooking water in rural communities.

Introduction

It is common knowledge that when people must go outside of their homes to obtain water, they will store it in a household water jar. In general water supply programs have emphasized protecting the water source while neglecting the HWJ. An incorrect assumption has often been made by water providers that the quality of water from the source was the quality of water consumed, when in fact testing of stored water has consistently shown it to be fecally contaminated, even when collected from a safe source (Feachem et al, 1978; Ryder et al, 1985; Lindskog & Lindskog, 1988). Thus there is an urgent need to include the HWJ in water supply programs. Targeting the HWJ may include changes in design, changes in behaviour, or disinfection. The present paper is the first report on the provision of safe water in rural villages based on disinfecting household drinking and cooking water in the HWJ.

After preliminary studies on well and HWJ quality, an appropriate chlorine dose was determined, village women trained, and on intervention trial carried out. The trial assessed the efficacy of HWJ chlorination on diarrheal morbidity and change in weight-for-height Z-score (WHZ) in children over a rainy season. The study sought to determine if household water chlorination could be recommended as a feasible, practical, cost effective means of providing safe water to rural residents of The Gambia. This work was carried out in 1990 and 1991.

Materials and methods

The study utilized small rural villages in one contiguous area of the Upper Baddibu District in the North Bank Division, The Gambia. All 22 communities in the study area whose primary source of drinking water came from open wells were included. In preliminary studies, the chemical, physical and microbiological quality of water from a sample of wells and HWJs was investigated, using standard testing procedures. The reaction of bleach (sodium hypochlorite) in the earthen HWJ was elucidated and organoleptic testing carried out to select an acceptable chlorine dose.

Village sensitization was performed in a series of meetings to explain the study objectives and methods; consent was given by each village. Next a sociodemographic census of all the compounds and households was performed. Each village was randomly assigned to either a control group or an intervention group. Comparability of the two groups of villages was assessed utilizing the sociodemographic census data. Each woman responsible for collecting water was given a 100 mL amber glass bottle, with childproof cap, and a plastic pasteur pipetree, clearly marked at the 1 mL point. Household bleach purchased locally was diluted to give a 2.5% (25,000 mg/L) chlorine solution. On collecting water, women were taught to put 2 mL into their standard 25 L plastic water collecting bucket, to give a chlorine dose of 2.0 mg/L. The water was then stirred and poured into their drinking and cooking HWJs. If the HWJ was more than half full on the following day, then an additional dose of 1 mL of the bleach solution was to be added. The training in all villages was the same. A placebo of sterile water was substituted in control villages.

The intervention study was carried out over a 20 week period, covering the 1991 rainy season. Fresh chlorine solution was delivered bi-weekly. Outcome indicators were the weight-for-height Z-score (WHZ) of children 6-24 months and 25-60 months, and their diarrhea incidence. Diarrhea was logged daily on a validated home-based recording form by the mothers, using their definition of diarrhea. Compliance was monitored by measuring chlorine residuals daily in a subset of villages, and spot testing for hydrogen sulfide producing bacteria and fecal coliforms in randomly selected HWJs in all villages. The study was planned to meet the methodological requirements of Blum and Feachem (1983) and the internal validation criteria of Esrey et al (1985) and Esrey and Habicht (1986).

Results

Water quality from open wells was acceptable for chlorination. Open well water was slightly acidic (pH 6.0-6.5), had a low buffering capacity, and highly corrosive. Ammonia was negligible and mean nitrate only half the WHO guidelines. The true colour (16 TCU) was at the WHO limit and turbidity averaged 23NTU, about five times above the WHO limit (WHO, 1984). Temperature was in the range 27° to 30°C. All well samples were contaminated and showed a mean fecal coliform level of 1871/100 mL (n=94); the one hour chlorine demand was 0.67 mg/L. Change in seasons did not significantly affect open well water variables. After 24-48 hours, water stored in the typically used HWJ contained a mean fecal coliform level of 3358/100 mL (n=52) in the rainy season and 1014/100 mL (n=32) in the dry season; 95% of HWJ samples were fecally contaminated. After reacting with the immediate chlorine demand, a dose of 2.0 mg/L dissipated slowly and continuously in the HWJ over a 24 hour period to an unsafe residual level below 0.2 mg/L. Factors affecting the rate of disappearance included the source well water quality, HWJ volume (i.e. reaction with earthen side), and chlorine demand introduced during storage. Under typical village conditions, 2.0 mg/L chlorine dose was sufficient to eliminate fecal coliforms within 30 minutes and to maintain microbiologically safe drinking water for 24 hours, even when high turbidity was present.

In the intervention study, an analysis of data aggregated by village showed that for both age groups, the mean diarrhea days per child was not significantly different between the two groups of villages. Although not significant at the 5% level of confidence, village mean weight gain was 27% higher in the villages chlorinating their HWJs for both age groups. Pre- and post-rain WHZ was compared in each village by a Paired t-test. The proportion of villages having a significantly lower post-rain WHZ was then compared using Fisher's Exact Test. In the younger age group fewer intervention villages had a significant decrease in WHZ at the 88% confidence level (1-tailed p-value = 0.117).

To increase statistical power, an analysis was performed of disaggregated data using each child as the unit of measure. The subset of malnourished children (i.e., WHZ < -2.0 standard deviations) was identified in each village. The proportion of pre-rain and post-rain malnourished children was compared in a 2X2 contingency table. Within the control group there was a significantly greater proportion of malnourished children at the post-rainy season survey as compared to the pre-rain survey in both age groups: 6-24 months p=.0002; 25-60 months p=.008. In contrast, there was not a significantly greater proportion of malnourished children in the post-rain survey for children drinking water from chlorinated HWJs. For both age groups the increase in the proportion of malnourished children in intervention villages was one-half that of control villages. Compliance monitoring revealed that on a daily basis, about 60% of intervention HWJs contained a chlorine residual or were microbiologically safe. The geometric mean fecal coliform level in 60 HWJs in 6 intervention villages was 178/100 mL, whereas in 30 HWJs in 6 control villages the level was 3020/100 mL.

Discussion

The study revealed that illiterate village women could properly dose their HWJs when given the proper training and supplies. Chlorinating HWJs at an organoleptically acceptable dose of 2.0mg/L provided safe drinking water for 24 hours. The evidence taken as a whole indicated that although a statistical association was not established between HWJ chlorination and improved village mean WHZ, a clear trend has been revealed in that direction. Chlorinating HWJs did prevent a significant proportion of children from becoming malnourished over the rainy season, particularly in younger children 6-24 months. The present study supports the thesis that water quality, not quantity, is more important to this younger age group.

There is evidence that chlorinating HWJs positively affected nutritional status, but no difference in diarrhea incidence was detected between study groups. This may be explained in two ways. First, the diarrhea surveillance instrument was not designed to reveal the intensity or severity of a diarrheal episode. For many enteric pathogens the severity of the diarrhea is related to the infecting dose. At the compliance level of the study, the dose of offending organisms may have been reduced such that diarrhea was present and thus recorded, but less severe. Or, the effect on nutritional status may have been independent of diarrhea (Stanton et al, 1988). Other factors which may have negatively affected the study results included:

1) other routes of transmission which masked the effect of HWJ chlorination,
2) a shorter than normal rainfall interval, in which over 90% of the rain fell in less than 12 weeks and a precipitation amount only two-thirds of the expected seasonal amount produced a lower than anticipated diarrhea incidence,
3) a lack of food availability, and
4) inadequate compliance level.

Although produced in Senegal and Mali, household bleach is generally imported from Europe and sold in town markets throughout The Gambia. To chlorinate one 50 L HWJ costs $0.002 per day, or $0.73 per year at current retail prices. Bulk purchasing for a HWJ disinfection program would likely reduce this cost by half. Costs of a HWJ chlorination program would mainly consist of purchase/distribution of disinfectant, educational programs on proper disinfection procedures, and related support and manpower requirements. Disinfectant could be distributed and sold using existing commercial outlets with a government-set price or within primary health care systems, utilizing the village health care worker. Multisectorial involvement with appropriate government departments, NGOs, and donor agencies could be encouraged and coordinated by the Department of Water Resources. Chlorination of HWJs is an appropriate strategy because household bleach is available and inexpensive, relies on existing open wells, involves the community as active participants (especially women), insures the ingestion of safe water by the entire household, and has a positive health impact on young children at risk of becoming malnourished over the rainy season. It is the hope of the author that agencies responsible for the provisions of safe water in developing countries will further explore HWJ disinfection.

Acknowledgements

This study was carried out in partial fulfilment of the Doctor of Science degree and was supported by funds from the International Board of the Southern Baptist Convention, Richmond, Virginia, U.S.A.

References

Blum, Deborah, & Feacham, R.G. (1983). *Measuring the impact of water supply and sanitation investments on diarrhoeal diseases; problems of methodology.* Int. J. Epidemiol. 12(3), 357-365.

Esrey, S.A., Habicht, J. & Butz, W.P. (1985). 'A methodology to review public health interventions: Results from nutrition supplementation and water and sanitation projects'. *Cornell International Nutrition Monograph Series* No. 15.

Esrey, S.A. & Habicht, J. (1986). *Epidemiologic evidence for health benefits from improved water and sanitation in developing countries.* Epid. Rev. 8. 117-128.

Feachem, Richard, Burns, E., Cairncross, S., Cronin, A., Cross, P., Curtis, D., Khan, M.K., Lamb, D., & Southall, H. (1978). *Water, Health and Development: An Interdisciplinary Evaluation.* London: Tri-Med Books Ltd.

Lindskog, R.V.M. & Lindskog, P.S. (1988). 'Bacteriological contamination of water in rural areas: an intervention study from Malawi'. J. Trop. Med. Hyg. 91, 107.

Ryder, R.W., Reeves, W.C., Singh, N., Hall, C.B., Kapikian, A.Z., Gomez, B., & Sack, R.B. (1985). 'The childhood health effects of an improved water supply system on a remote Panamanian island'. Am. J. Trop. Med. Hyg. 34(5), 921-924.

Stanton, B.F. Clemens, J.D., Khair, T. (1988), 'Educational intervention for altering water sanitation behavior to reduce childhood diarrhea in urban Bangladesh: impact on nutritional status'. Am. J. Clin. Nutr. 48, 1166-1172.

World Health Organization, *Guidelines for Drinking Water Quality.* Vol. 2. 1984. Health criteria and other supporting information. Geneva: World Health Organization.

Domestic solar disinfection of potable water

Michael D Smith

BACTERIA WHICH ARE present in water may be killed using a variety of different techniques; such as addition of certain oxidising chemicals, or exposure to heat or radiation. The selection of these disinfection options depends on the volume of water to be supplied, finances and materials available, and the initial quality of the water. Disinfection of domestic water supplies, on a household basis and at low cost, presents particular difficulties.

Experimental work undertaken in Beirut indicated that coliforms present in water stored in transparent water containers were killed within a few hours when the containers were exposed to sunlight (UNICEF, 1984). Tests on samples of water containing cultured pathogenic organisms and stored in 300 ml round pyrex flasks and exposed to intense sunlight showed that the time required for complete destruction of organisms ranged from 15 minutes *(P aerugenosa)* to 90 minutes *(S paratyphi B)*. Under these conditions coliform bacteria were destroyed in 80 minutes. The results suggested that bacteria were killed by exposure to solar radiation in the near-ultraviolet region (315 - 400 nanometres) of the light spectrum. The researchers suggested that their findings had applications for disinfecting small quantities of water for both drinking and preparation of Oral Rehydration Therapy (ORT) solutions.

The Beirut experiments operated on a batch process, in that fixed volumes of water were exposed to sunlight. The time taken for all bacteria to be killed depended upon such variables as the sunlight intensity, thickness of the container walls, container material and colour, initial water quality and container dimensions. In 1990, work started at WEDC (Loughborough University) to investigate whether solar radiation could be used to disinfect water on a continuous (steady flow) basis. The experimental work has been conducted over three years as project work by individual participants on MSc programmes.

It is well-known that certain wavelengths of ultraviolet light are very effective at killing bacteria; the wavelengths from 200 to 290 nm (nanometres) being most effective. The optimum wavelength varies between organisms, but is generally around 260 nm. Light having wavelengths from 370 nm (in the near ultra-violet) to 500 nm (in the blue-green region of the visible spectrum) also have a weaker ability to kill bacteria. For the purposes of the experimental work at WEDC it was assumed initially that solar radiation was responsible for the death of bacteria.

The most favourable locations for solar applications are those countries lying between 15° and 35° North and South, where there is greatest solar radiation intensity, and limited cloud cover and rainfall. The UK, which lies between latitudes 50 and 59°, is not a favourable country for undertaking research into solar disinfection, yet work was possible during periods of fine weather in summer months.

Experimental programme at Loughborough

For the experimental procedures adopted, water flowed slowly from one storage container into another. While flowing, the water was protected from external pollution but was exposed to sunlight. The water samples used consisted of settled sewage, from a local wastewater treatment plant, which were diluted to simulate untreated water. A peristaltic pump was used to maintain a slow but constant flow rate for the transfer of water between containers, so that the water was exposed to solar radiation for as long as possible; for periods of between 2 and 30 minutes. During the tests, solar radiation intensity was measured using an energy sensor, and water samples for bacteriological analysis were taken from both water containers.

Experimental equipment, and especially the panel through which water flowed between storage containers, was modified and improved over the three experimental periods. Efforts were made to keep the equipment both cheap and simple.

During the first year the panel consisted of an inclined metal tray, covered by glass. The water did not, however, flow in a thin film as was hoped, condensation caused the glass to become misted, and the glass did not transmit certain wavelengths of solar radiation (Mozah, 1990).

For the second year, several lengths of Polyvinylidene (PVDF) tubing were placed parallel to one another on an inclined tray and connected together to create a total flow path 9 metres long. PVDF tubing was selected because it is translucent to visible light; and transmits, and is inert to, UV radiation having a wavelength above 150 nm. The tubing used was of 10 mm outside diameter, with a wall thickness of 1 mm (Hussain, 1991).

For the third year, flow was divided between two lengths of PVDF tubing, each having a length of 3 metres. One length of tubing was covered in opaque black tape; which had the effect of shielding the water from direct sunlight, but increasing the temperature by absorbtion of radiation (Addy, 1992).

Results and conclusions

Selected results obtained from experimental programmes are shown in Table 1, below.

Results obtained from experimental work at WEDC to date suggest that bacterial death may result both from exposure to solar radiation and an increase in temperature. The relative importance of each effect has yet to be ascertained, and further study is required in order to understand the conditions necessary for reliable solar disinfection of potable water.

Experimental work in this subject, using both batch and continuous flow processes, is continuing in several countries worldwide (IDRC, 1988). Experimental equipment being used varies considerably in both cost and complexity, and experimental results indicate that bacterial kill depends both on the radiation intensity and exposure time.

References

Acra, A.; Rafford, Z.; Karahagopian, Y. (1984) *Solar Disinfection of Drinking Water and Oral Rehydration Solutions, Guidelines for Household application in Developing Countries*. UNICEF, Amman, Jordan.

Addy, J.K. (1992) *The effect of ultraviolet radiation and temperature in solar disinfection of water*. MSc Thesis. WEDC, Loughborough University of Technology, U.K.

Hussain, M.Y.M. (1991) *Solar disinfection of water*. MSc Thesis. WEDC, Loughborough University of Technology, U.K.

IDRC (1988) 'Solar Water Disinfection'. IDRC-MR231e. Proceedings of a workshop held at the Brace Research Institute, Montreal, Quebec, Canada, 15-17 August 1988. IDRC (International Development Research Centre) Canada.

Mozah, J.A. (1990) *Disinfection of water by exposure to sunlight*. MSc Thesis. WEDC, Loughborough University of Technology, U.K.

Table 1. Summary of results of solar disinfection studies
(only highest bacterial kill rates are shown)

PLACE	DATE	% REDUCTION IN COLIFORM BACTERIA		EXPOSURE TIME (mins)	SOLAR RADIATION INTENSITY (W/m^2)	TYPE OF MATERIAL USED FOR WATER EXPOSURE
		TOTAL	FAECAL			
BEIRUT	1979	82	99.99	60	Not known	Glass flasks
WEDC	1990	78	88	2	Not known	Metal tray covered with glass sheet.
WEDC	1991	94	86	30	700	Polyvinylidene fluoride tube
WEDC	1992	Not known	63	5	226	Polyvinylidene fluoride tube
		Not known	34	5	226	Wrapped Polyvinylidene fluoride tube

Moringa oleifera at pilot/full scale

J P Sutherland, G K Folkard, M A Mtawali and W D Grant

CRUSHED SEEDS OF the tree Moringa oleifera Lam. (M.oleifera) are a viable replacement coagulant for proprietary chemicals such as aluminium sulphate (alum) in developing countries. The tree is a multi-provider that grows pantropically and its distribution in Africa and various vernacular names are noted.

This paper presents results from pilot scale treatment trials carried out at Thyolo in southern Malawi early in 1993. The pilot works utilised for the study, situated on the site of an existing Ministry of Works operated treatment plant, comprises a header/flash mixing tank, gravel bed flocculators, plain horizontal flow sedimentation tank and a rapid gravity filter. The system has a nominal flow rate of 1 m³/hr. Inlet raw water turbidities were maintained at around 400 NTU throughout the six week study. Over 90% removal of turbidity was achieved by effective floc formation and sedimentation. Floc carry over was susequently removed in the filter producing a final water consistently less than the WHO guideline value of 5 NTU.

Results from full scale trials on the existing Ministry of Works operated treatment plant are also presented. Alum, the normal coagulant used, was replaced with M.oleifera seed solution and comparable performance was observed. This is the first time that M.oleifera has been used as a primary coagulant at this scale (flow rate 16 m³/hr).

Background

Since 1986 the Environmental Engineering Group at the University of Leicester have been examining crushed seed powder of the tree Moringa oleifera Lam. (M.oleifera) as potential full or partial replacements for proprietary chemical coagulants in the treatment of surface waters in developing countries. M.oleifera is a native of the sub-himalayan tracts of N.W.India, Pakistan and Afghanistan and indigenous to many areas of Africa, South America and Asia. Various vernacular terms for the tree associated with Africa include (Jahn, 1986):

Nigeria - Adagba Malero
Burkina Faso - Argentiga/La-Banyu
Malawi - Chamwamba/Kangaluni/Sangoa
Ghana - Ewe Babatsi/Ewe Yevuti
Kenya - Mborongi
Tanzania - Mlonge/Mronge
Gambia - Neberdayo

The seed pods are allowed to dry naturally on the tree prior to harvesting. The seeds are easily shelled, crushed and sieved using traditional techniques employed for

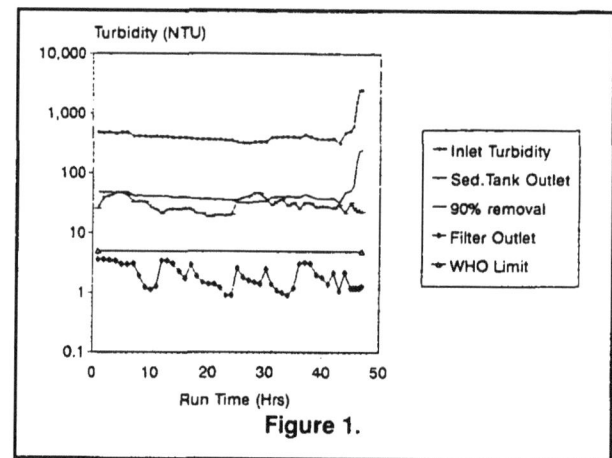

Figure 1.

the production of maize flour. Dosing solutions are generally prepared as 1-3% solutions. The full scale trial reported was dosed with a 5% seed solution due to the limited capacity of the peristaltic dosing pumps available. The crushed seed powder, when mixed with water, yields water soluble proteins that possess a net positive charge. The solution acts as a natural cationic polyelectrolyte during treatment (Sutherland, Folkard and Grant, 1990).

In addition to the use of the seeds as a coagulant, the seeds, pods, flowers and leaves are used for a wide range of other purposes from foodstuffs to traditional medicine to oil production (Jahn 1986).

Pilot plant

The pilot plant consists of a header/mixing tank, gravel bed flocculation, plain horizontal sedimentation and rapid gravity filtration. Nominal flow rate through the plant is 1 m3/hr. Details of the works have been reported previously (Folkard, Sutherland and Grant, 1993). Seeds for the study had been collected and processed by the Forestry Research Institute of Malawi (FRIM). As previously stated raw water turbidities were maintained at around 400 NTU throughout the field study. Figure 1 shows a composite of the runs carried out. It can be seen that solids removal following flocculation and sedimentation is consistently in excess of 90% with filtered water quality remaining below the WHO guideline value of 5 NTU. Seed dose ranged from 75 - 250 mg/l dependent on the initial raw water turbidity. Figure 2 shows results obtained from a single 7 hour run.

Full scale trial

It had originally been planned to carry out full scale trials over a two week period. However, due to a lack of seed only a single 6 hour run using the seed was possible. The

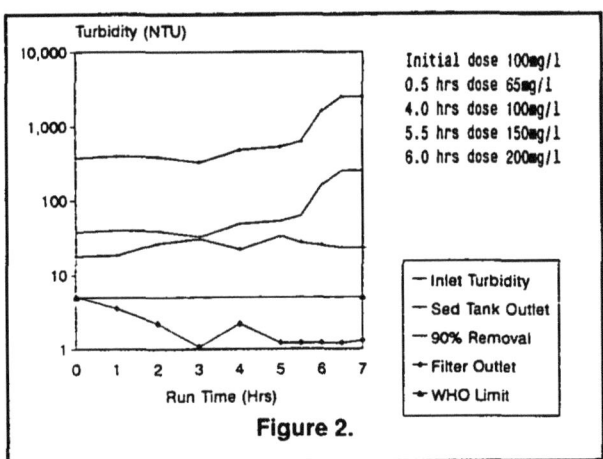

Figure 2.

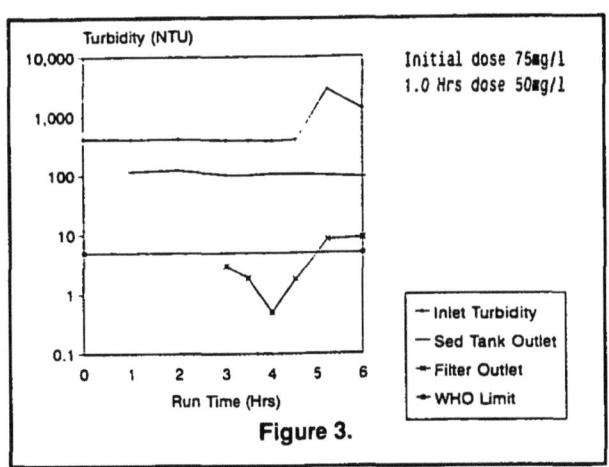

Figure 3.

main works at Thyolo comprises two reactor clarifiers for primary solids removal followed by rapid gravity filters. Since installation in the 1970's both clarifiers have fallen into a state of disrepair. The impellers used to control the flow regime and to provide efficient mixing during flocculation are no longer operational. In addition, effective desludging can only be carried out manually, the time between desludging being judged by the operator as the sludge level indicators are inoperative. This can result in excessive sludge build up further reducing the efficiency of the works.

Due to the limited quantity of seed available for the trial only one of the clarifiers was used. The flow rate through the clarifier was determined to be 16 m^3/hr. The works was initially monitored with aluminium sulphate (alum) as the coagulant. The alum was then replaced with a suspension of M.oleifera. The coagulant was dosed via two peristaltic pumps into the nappe of the discharge from a 'V' notch weir in the raw water inlet box.

Figure 3 shows results obtained using alum as the coagulant. Figure 4 shows results obtained when alum was replaced with the seed solution. Comparing the two it can be seen that there is no deterioration in performance. Equivalent removals are achieved using the seed although at a higher dose than alum. The inherent inefficiency of the works is evident from the clarifier output turbidity. Excessive floc carry over was observed for both coagulants.

Conclusions

The pilot plant trials have demonstrated the effectiveness of the seed treatment for the clarification of high turbidity waters at 1 m^3/hr.

The full scale trial has demonstrated the viability of using the seed suspensions at 16 m^3/hr with equivalent performance to that of alum being achieved. Although higher seed doses were required studies are currently under way to improve processing techniques with the aim of producing a finer seed powder increasing the amount of active coagulant released.

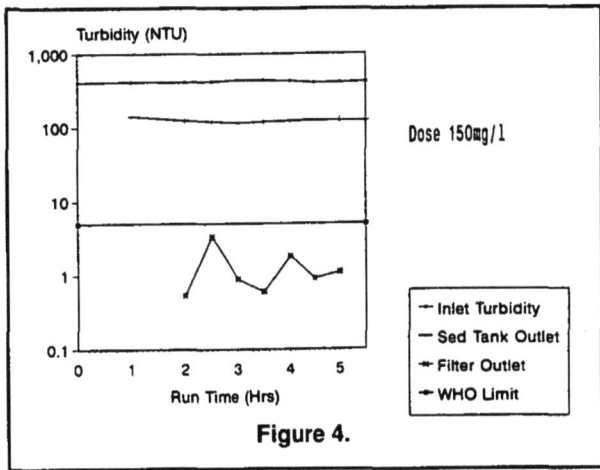

Figure 4.

Future work

Further full scale trials in Malawi are planned for January 1994. It is hoped that during this period demonstrations for interested parties from other developing countries will be arranged. Those interested in participating may receive further information from the authors.

Acknowledgements

The authors gratefully acknowledge the financial support of the Overseas Development Administration of the British Government and the assistance of personnel from the collaborating agencies in Malawi.

References

JAHN, S.A.A., 1986. Proper use of African natural coagulants for rural water supplies. GTZ, Eschborn, Germany

SUTHERLAND, J.P., FOLKARD, G.K. and GRANT, W.D., 1990. Natural coagulants for appropriate water treatment: a novel approach, Waterlines, April, 8, 4, pp30-32.

FOLKARD, G.K., SUTHERLAND, J.P. and GRANT, W.D., 1993. Natural coagulants at pilot scale; In: Pickford, J. ed. Water, Environment and Management: Proc. of the 18th WEDC Conference, Kathmandu, Nepal, 30 Aug.-3 Sept.1992. Loughborough University, pp51-54.

Water treatment in Northern Ghana

Martin Wegelin and Kolly Dorcoo

THE GHANA WATER AND SEWERAGE CORPORATION (GWSC) is in the process of rehabilitating water supply systems in the Northern Region of Ghana with external financial and technical support. This programme foresees also the rehabilitation and extension of water treatment plants of 13 towns whose projected population ranges between 5,000 and 73,000 inhabitants. Nine towns use as raw water source impounded reservoirs fed by non-perennial surface runoff. The other 4 towns draw their raw water from rivers.

The three existing conventional plants whose treatment processes comprise flocculation, sedimentation, rapid sand filtration, and chlorination, are only partly operational. The sedimentation tanks at one treatment plant have serious design deficiencies, and the construction of the rapid sand filters have never been completed. Consequently, the alum flocs are washed straight into the clear water tank and, thereafter, pumped into the distribution network. At the other two sites, the filter medium in the rapid sand filter is missing. Hence, only partly treated water is distributed to the consumers. The other towns have to rely on package plants, which include a 3-tank system composed of sedimentation tank, rapid sand filter pressure tank and clear water tank. The package plants are installed as closed system in the raw water line in front of the high level tank. Alum, lime and bleaching powder are added to the water by chemical dosers. The package plants are not functioning well or are out of operation. Difficulties with the supply of chemicals, partly broken dosage equipment or uncontrolled operation of the package plants result in an unreliable water quality. After less than one year of operation, a package plant which had been submitted to major rehabilitation was only partly functional. Maintenance of the installations and dependency on the import of foreign material are serious problems faced by GWSC.

Treatment options

For the large water supply schemes, the preliminary design report (1) of the rehabilitation project proposes to either extend or construct new conventional treatment plants with flocculation, sedimentation, rapid sand filtration, and chlorination as treatment processes. The existing package plants in smaller water supply schemes should be rehabilitated and additional units running in parallel should be added if necessary. Furthermore, construction of one slow sand filter plant was planned for one water supply scheme.

As a result of the experienced difficulties with conventional water treatment plants and with the installed package plants, GWSC embarked on field tests with small slow sand filter units. Since slow sand filtration has, so far, not been applied in the project area, field tests were carried out at two sites to gain practical experience with this treatment process.

As slow sand filtration is a simple and reliable water treatment process making greatest use of locally available resources, it is considered an appropriate technology for developing countries. However, reasonable filter operation is only possible with raw water of low turbidity. Pretreatment of surface water is generally necessary to achieve slow sand filter runs of 1-2 months or more. Sedimentation combined with chemical flocculation is applied in conventional water treatment for the reduction of turbidity and separation of solid matter. Chemical flocculation, however, is a rather sensitive and unstable process difficult to control. Flocs often escape the sedimentation tanks and rapidly clog the subsequent slow sand filters. Chemical flocculation should therefore never directly precede slow sand filtration without the use of another intermediate pretreatment step.

Filtration is an alternative and efficient pretreatment process for the removal of solid matter. Roughing filters use coarse filter material, do not require sophisticated mechanical equipment and are operated at low filtration rates without the addition of chemicals. Roughing filters, which were developed over the last decade, are now used world-wide as efficient and appropriate pretreatment process prior to slow sand filtration (2). The layout of a treatment scheme adequate for rural water supplies is illustrated in Fig. 1.

Pilot plant tests

Motivated by the simplicity of the horizontal-flow roughing filter design, the GWSC project team tested this pretreatment process in combination with slow sand filtration at two sites. Preliminary and valuable practical experience (3) was made with the installation of a pilot plant and the carrying out of field tests at Damongo, a district town supplied by an impounded reservoir. Although the raw water turbidity was found to be low (20 - 30 NTU), iron problems from both natural (dissolved iron concentration in the raw water abstracted from the reservoir's bottom ranged between 0.4 and 1.8 mg/l) and man-made sources (iron release by laterite filter material and rusty steel drums) impaired the field tests. The growth of iron bacteria in the roughing and slow sand filters hindered a satisfactory filter performance. The gravel of the roughing filter units and the rusty steel drum used as slow sand filter unit were replaced by adequate filter material and by an asbestos cement pipe respectively. Reasonable slow sand filter runs of 70 days were then achieved with the pilot plant. For the full-scale treatment plant, however, abstraction of the raw water at the reservoir's surface by a floating intake and separation of the dissolved iron by efficient aeration (cascades) and sedimentation prior to filtration is recommended.

The lessons learnt at Damongo were applied in the design of the second pilot plant constructed at Salaga where river water is used as raw water source. An aeration/precipitation unit consisting of a tray aerator and a sedimentation tank was constructed prior to the two horizontal-flow roughing filter units built in concrete block work and four slow sand filter units installed in asbestos cement pipes. The filter units were operated at filtration rates of 0.75 and 1.5 m/h, and of 0.1 and 0.2 m/h respectively. According to the field tests (4) carried out in the dry season, the relatively low raw water turbidity ranging between 15 and 30 NTU was reduced in the pilot plant to generally less than 5 NTU (5.8 - 1.3 NTU). However, during the rainy season the raw water turbidity increased to 350 NTU but was then reduced to 150 NTU by the horizontal roughing filter, and to 120 - 140 NTU by the slow sand filter units. Fine colloidal particles in the treated water were the cause for the low removal efficiency in the slow sand filters. Bacteriological analyses, however, revealed a satisfactory and acceptable water quality. After 6 months of operation, the headlosses in the roughing filters increased to only 1 cm and the four slow sand filters, except for one unit, never had to be cleaned during this period. Apart from some aesthetic water quality deficiencies caused by removal difficulties of the colloidal particles, the field tests proved the operational viability of the tested treatment process. Other field tests in the project area carried out with a horizontal-flow roughing filter unit by the Village Water Reservoirs Project revealed that turbidity could generally be reduced from approx. 500 NTU to 170 NTU (5). However, it was observed that the cyclopes acting as host for the guinea worm larvae were not fully separated by single filtration. Complete cyclopes removal would therefore require the addition of slow sand filters.

Two full-scale treatment plants consisting of horizontal-flow and slow sand filters were constructed in Ghana at Mafi Kumase and Katamanso and complement the experience made with the above-mentioned pilot plants. In Mafi Kumase, the guinea worm disease could be eradicated and the slow sand filters reveal with a running time of 6 months an excellent performance.

Alternative pretreatment options

In Ghana, only horizontal-flow roughing filters have so far been used as prefilters. This filter type, however, consisting of a relatively large filter bed of usually 5-7m length, is generally used for the pretreatment of highly turbid water. For the pretreatment of raw water with low and moderate turbidity, smaller filter structures can be applied. Alternative prefilters types such as intake and dynamic filters, as well as upflow roughing filters are illustrated in Fig. 2. Hence, the raw water of impounded reservoirs will generally have to undergo pretreatment in small prefilters (intake or dynamic filters) and in upflow roughing filters.

Cost comparison

Construction and operation costs of a new treatment plant at Damango with a design capacity of 1110 m3/d and a projected population of 16,600 inhabitants for the year 2000 were estimated on the basis of preliminary designs (1, 6) for the following treatment plant options:

a) conventional treatment plant comprising flocculation tanks (Td 30 min.), sedimentation tanks (Td 4 hrs), rapid sand filters (vF 6 m/hr), and dosage of alum, chlorine and lime
b) alternative treatment plant comprising horizontal-flow roughing filters (vF 1.1 m/hr, Ltot 10 m (!)), slow sand filters (vF 0.11 m/hr), and dosage of chlorine
c) modified alternative treatment plant comprising aeration towers, sedimentation tanks (Td 30 min.), upflow roughing filters (VF 0.8 m/hr, Ltot 1 m), slow sand filters (VF 0.15 m/hr), and dosage of chlorine
(Td detention time, vF filtration rate, Ltot total filter length)

Designs a) and b) are based on design guidelines recommended in the literature, whereas design c) applies the recommendations made on the basis of the pilot test results (3, 6). The results suggest that substantial cost reductions can be achieved especially with the type of pretreatment since low turbidity records enable a shift from large horizontal-flow to much smaller upflow roughing filters. A relative cost comparison is given in Table 1. In general, the foreign currency demand of the alternative options is about half of that required by the conventional treatment system. With respect to the construction costs, the modified alternative treatment plant (c) is competitive to the conventional treatment plant (a) which, however, requires almost half of the investment costs in foreign currency for its installation. As regards to the operation costs, the roughing filter options (b, c) are significantly cheaper due to the lower chemical demand. Finally, the specific per capita construction and operation costs for the treatment facilities are also indicated (price basis 1992, 1 US $ = 400 Cedis).

Implementation

Rehabilitation of the water supply schemes should be demand-oriented rather then supply-oriented and geared towards community self-managed systems. Therefore, community meetings were held and "willingness-to-pay" surveys carried out in order to determine and enhance communal managerial skills and assess its financial capacity necessary for the implementation and successful operation of the water projects. Additional field tests with pilot plants are proposed in order to assess the performance of the treatment schemes and to develop economic designs of the full-scale plants. The pilot plants could also be used as demonstration units for the first communities to be supplied with roughing and slow sand filtered water. The water quality produced by the pilot plants, especially its appearance, taste and odour will be judged by the consumers whereas the

Table 1. Cost comparison for different treatment options

treatment scheme	construction costs			operation costs			specific costs	
	local	foreign	total	local	foreign	total	constr.	operation
flocculation, sedimentation rapid sand filtr., chlorination	54%	46%	100%	21%	73%	100%	14.4$/c	1.0$/c.yr
horizontal-flow roughing filtr., slow sand filtr., chlorination	160%	18%	178%	32%	37%	69%	25.5$/c	0.7$/c.yr
aeration/sed., upflow rough. filtr. slow sand filtr., chlorination	66%	18%	84%	25%	37%	62%	12.1$/c	0.6$/c.yr

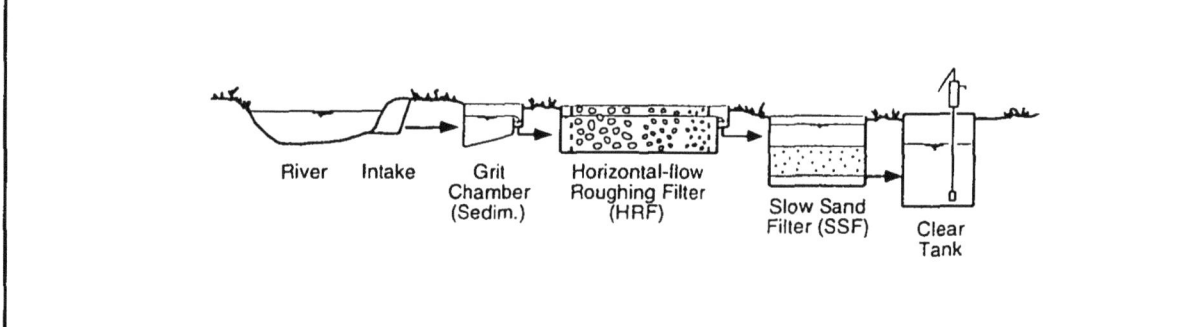

Figure 1. Layout of a self-reliant water treatment plant

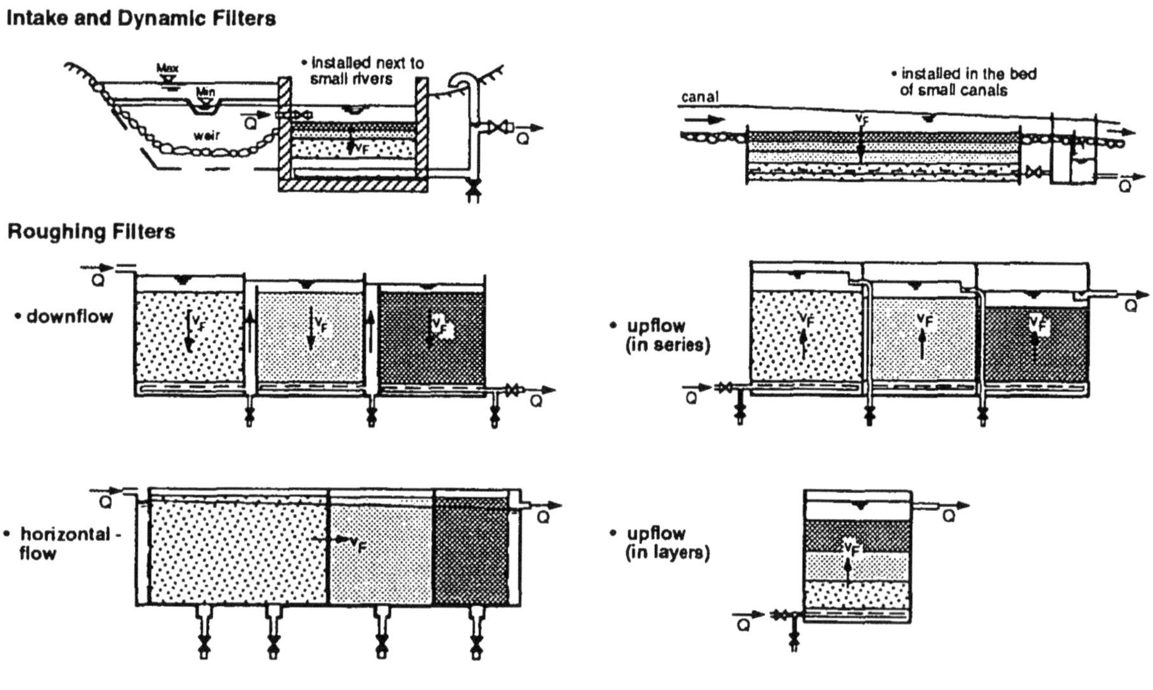

Figure 2. Layout of different roughing filters

operators will be able to familiarise themselves with the type of treatment process and maintenance work they will have to carry out. Full-scale treatment plants can serve as demonstration units to other communities once the treatment process is introduced in the region.

If water supply projects are fully implemented in one phase they require considerable investment within a short period. This approach is neither flexible with regard to future modifications, nor does it make allowances for the rather slow development and consolidation of community management. Involvement of the community in the design, decision and participation process is limited and can therefore seriously endanger the sustainability of water supply projects. Phased implementation, however, keeps pace with community development and is thus a solution to the above-mentioned problems. Such an approach, which will gradually increase the service level of the water supply, will develop the interest, management capacity and financial strength of the community. A phased implementation might include the following steps:

- pilot plants to demonstrate treatment processes
- full-scale treatment plant supplying a clear water well furnished with handpumps
- construction or rehabilitation of the distribution network equipped with public standpipes
- extension of the distribution network and possible installation of private connections
- extension and, if required, upgrading of the treatment facilities

Simple and reliable treatment processes in rural areas are essential for sustainable water supply schemes. However, the construction of a water supply scheme is also a process during which building of local management capacity is equally important.

References

(1) Preliminary design report, Wardrop, July 1991
(2) The decade of roughing filters, M. Wegelin at al, AQUA, No. 5/1991
(3) Damongo slow sand filtration pilot project, GWSC, March 1992
(4) Horizontal-flow roughing and slow sand filtration pilot plants experience in the Northern Region, Ghana, Ch. Berhoh, International workshop on roughing filters for water treatment, Zurich, June 1992
(5) Horizontal roughing filtration, Experimental Report, J. Addy and A. Ligtenberg, Village Water Reservoirs Project, July 1991
(6) Potential of slow sand filters in the northern region of Ghana, M. Wegelin and K. Dorcoo, Field report No. 7, Wardrop. April 1992

www.ingramcontent.com/pod-product-compliance
Ingram Content Group UK Ltd.
Pitfield, Milton Keynes, MK11 3LW, UK
UKHW050226150426
5217IPUK00023B/1667